SHACKLETON

A BEGINNER'S GUIDE

CHRISTOPHER EDGE

Hodder & Stoughton

A MEMBER OF THE HODDER HEADLINE GROUP

ACKNOWLEDGEMENTS

I wish to thank the staff at the Scott Polar Research Institute in Cambridge, especially archivist Robert Headland.

DEDICATION

To Chrissie for all her support and encouragement.

Orders: please contact Bookpoint Ltd, 130 Milton Park, Abingdon, Oxon OX14 4SB. Telephone: (44) 01235 827720, Fax: (44) 01235 400454. Lines are open from 9.00–6.00, Monday to Saturday, with a 24-hour message answering service. Email address: orders@bookpoint.co.uk

British Library Cataloguing in Publication Data
A catalogue record for this title is available from The British Library

ISBN 0 340 84645 3

First published 2002
Impression number 10 9 8 7 6 5 4 3 2 1
Year 2007 2006 2005 2004 2003 2002

Copyright © 2002 Christopher Edge

Figures 1.2 and 6.1 Copyright © 2002 Christina Edge

All rights reserved. No part of this publication may be reproduced or transmitted in any form or by any means, electronic or mechanical, including photocopy, recording, or any information storage and retrieval system, without permission in writing from the publisher or under licence from the Copyright Licensing Agency Limited. Further details of such licences (for reprographic reproduction) may be obtained from the Copyright Licensing Agency Limited, of 90 Tottenham Court Road, London W1P 9HE.

Cover photo from Sean Sexton Collection/Corbis.
Typeset by Transet Limited, Coventry, England.
Printed in Great Britain for Hodder & Stoughton Educational, a division of Hodder Headline Plc, 338 Euston Road, London NW1 3BH by Cox & Wyman, Reading, Berks.

CONTENTS

INTRODUCTION 1
A note on the text 2

CHAPTER 1: ANTARCTICA – THE LAST WILDERNESS ON EARTH 3
Extreme conditions 4
Wildlife 5
The discovery of Antarctica 7
The voyages of Ross 9
Summary 11

CHAPTER 2: IN SEARCH OF ADVENTURE 12
Learning the ropes 13
Scott and the *Discovery* expedition 15
The voyage to Antarctica 17
Summary 20

CHAPTER 3: THE STRUGGLE FOR SUPREMACY 21
The South Polar Times 23
The journey south 24
Starvation and scurvy – a race for life 27
Invalided home 28
Summary 30

CHAPTER 4: THE STRENGTH TO TURN BACK	31
The *Nimrod* expedition	32
A broken promise	35
The Antarctic winter	36
The road to the Pole	38
Furthest south and failure	40
Forty hours without food	41
Summary	43
CHAPTER 5: *ENDURANCE*	44
Plans to cross Antarctica	45
The *Endurance*	46
Into the Weddell Sea	48
Trapped	50
Abandon ship	52
Summary	56
CHAPTER 6: LIFE ON THE ICE	57
Ocean Camp	59
The *Endurance* sinks	59
A mutiny averted	60
Launch of the lifeboats	62
Elephant Island	63
Summary	65
CHAPTER 7: INTO THE UNKNOWN	66
Life on Elephant Island	66
The stormiest seas on Earth	67
Beyond the limits of endurance	69
South Georgia, but not safety	71

The crossing of South Georgia 73
Stromness Bay 75
Summary 76

CHAPTER 8: SHACKLETON'S LEGACY 77
Rescue from Elephant Island 77
The *Quest* expedition 80
The modern age of exploration 81
In Shackleton's footsteps 83
Summary 84

GLOSSARY 85

FURTHER READING 87

USEFUL ADDRESSES 89

INDEX 90

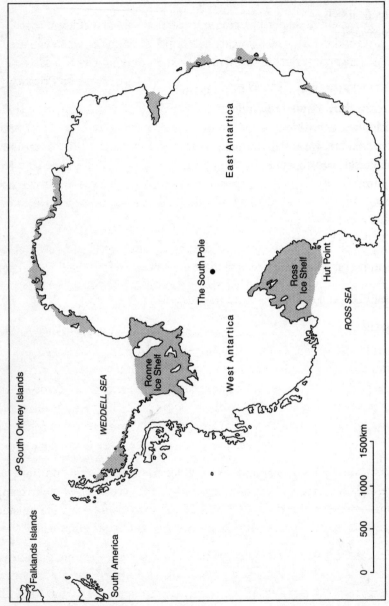

Figure 1.1 Antarctica and its surrounding seas

INTRODUCTION

> 'For scientific leadership, give me Scott, for swift and efficient travel, Amundsen. But when you are in a hopeless situation, when you are seeing no way out, get down on your knees and pray for Shackleton.'
> Sir Raymond Priestley

In the history of the Antarctic, the names of three explorers shine brightly: Amundsen, Scott and Shackleton. Amundsen and Scott both reached the South Pole, an achievement of such magnitude that ensured that their names would never be forgotten. Shackleton's name endures, however, because of the remarkable nature of his exploits, the bravery of his actions and the way that he ensured the survival of the men he led in the bleakest of times.

In August 1914, in the wake of Amundsen and Scott's conquest of the South Pole, Shackleton set off on an expedition to lead the first crossing of the Antarctic continent. As the First World War began to unfold in Europe, Shackleton and his crew headed for the Antarctic seas.

The journey Shackleton had embarked upon would become legendary, but it was not the story of the first Antarctic crossing. In January 1915, with Shackleton's ship the *Endurance* in sight of the Antarctic coastline, the pack-ice on the sea closed around the ship on all sides. Trapped, the ship was gradually crushed by the pressure of the ice pack. As their ship was claimed by the ice and sea, Shackleton and his men were forced to abandon the *Endurance* and take their chances on the floating pack-ice.

They were stranded, hundreds of miles from safety with no hope of rescue. The ice on which they stood was a thin crust above the freezing Antarctic seas; they had no means of communicating with the outside world and only limited supplies of food and essential equipment.

Shackleton's struggle to keep his men alive through the long months of the Antarctic winter, as they drifted away from land and safety, was a fierce test of his leadership and resilience. The journey he would lead his men on would take them through the depths of suffering and force them to face the most arduous of challenges. They would push themselves to the edge of physical endurance, face the psychological torments of isolation and loneliness, and journey into the heart of the unknown. The survival of the entire expedition crew, with not one single life lost, meant that Shackleton's name became a byword for heroic leadership and bravery.

Shackleton's achievements are even more incredible when seen from an historical perspective. The world of the early twentieth century was a very different place from the world we live in today. Communication was by letter and telegraph, not e-mail and satellite as it is today. The modern explorer can rely on the use of radios, GPS technology, and the support of emergency back-up teams. Shackleton and his men were beyond the eyes and ears of the world, completely cut off from civilization.

As well as belonging to the pioneering age of polar exploration, Shackleton was in many ways a particularly modern type of celebrity. At a time when most exploration was funded by national governments and scientific institutions, Shackleton resorted to self-promotion, newspaper advertisements and sponsorship from wealthy businessmen to raise funds in order to realize his dreams of adventure and discovery. Shackleton was also one of the first explorers to realize the commercial appeal of his expeditions and encouraged people to invest in his ventures with the promise of future royalties from the sale of the books and films of the expeditions.

The continent of Antarctica at the turn of the twentieth century was a gaping void in people's knowledge of the world. They looked to Shackleton and men like him to brave the perilous Antarctic wastelands and bring back news from the bottom of the world.

A NOTE ON THE TEXT
You will see that key terms and unfamiliar words are set in **bold** text. The keywords are defined and explained in the Glossary to be found at the back of the book.

Antarctica – The Last Wilderness on Earth

Antarctica is a place of extremes, the coldest, driest, windiest and highest of the seven continents. Completely surrounded by the Southern Ocean it exists in splendid isolation, with its ice-packed seas protecting it from scrutiny.

It is the fifth largest of the seven continents, eclipsing Europe and Australia in size. The Antarctic continent is divided into two distinct regions, East Antarctica and West Antarctica, bisected by the Transantarctic mountains, a mountain range that stretches for 2900 kilometres. Antarctica is a land of contradictions. It is a desert and the driest continent on earth, yet it contains 90 per cent of the world's ice and 70 per cent of the world's fresh water (in the form of ice). If all of the ice in Antarctica were to melt, sea levels around the world would rise by over 60 metres, cities would drown and chaos ensue. Antarctica's average **elevation** is 2250 metres above sea level and the continent's highest point, Vinson Massif, is 4897 metres.

Antarctica is covered by a massive sheet of ice, the Antarctic Ice Sheet, which extends over the continent and out over the sea for 2.6 million square kilometres in the summer, growing to 18.5 million square kilometres in the winter, doubling the area of the continent. This ice sheet is on average 2.7 kilometres thick and has been formed by layers of ice and snow building up over millions of years. This has been a slow process as Antarctica's average annual **precipitation** is less than five centimetres.

> **KEYWORDS**
> Elevation: height
> Precipitation: rain or snowfall.

The Antarctic Ice Sheet is so thick that entire mountain ranges are completely buried beneath it. Under the pressure created by its own weight the ice flows outwards from the high central interior of

Antarctica, down slopes in the form of **glaciers**. The largest glacier in the world, the Lambert Glacier, is found in Antarctica. It is 40 kilometres wide and flows for 400 kilometres through East Antarctica before emptying into the Emery Ice Sheet.

When many of the great Antarctic glaciers reach the coast they converge and join together to form the immense floating ice shelves that protect 30 per cent of the 17,500 kilometres of coastline. The largest of the Antarctic ice shelves is the Ross Ice Shelf, which is equal in size to France. These floating ice shelves are not stable masses of ice and sometimes whole sections **calve** away into the Antarctic seas.

> **KEY FACT**
>
> Like a river, a glacier flows at different speeds. The fastest ice is found at the centre and on the surface, whilst the slowest ice is at the margins of the glacier and at its base. The ice on the surface of a glacier is brittle and develops crevasses – deep cracks or holes in the ice.

EXTREME CONDITIONS

> **KEYWORDS**
>
> Calve: throw off masses of ice.
>
> Vostok Station: a research base at the Geomagnetic Pole.

Antarctica is renowned for its extreme coldness. The lowest temperature ever recorded on the earth's surface was measured at −89.6° C, and recorded at Russia's **Vostok Station** in Antarctica on 21 July, 1983. There are several reasons for Antarctica's extreme cold: its polar location, the continent's height, and the ice sheet which reflects 80 per cent of the sun's radiation back into space. In the interior of Antarctica, average temperatures range from −40° C to −70° C at the height of winter, and between −15° C to −35° C during the warmest month. Along the coast, temperatures increase and range from −15° C to −32° C in the winter and +5° C to −5° C in the summer. However, these temperatures do not take into account the effect of the strong Antarctic winds which can reach speeds of up to 320 kilometres per hour. These high speed winds severely increase the chill factor for intrepid explorers.

Antarctica's weather is very unpredictable: a clear day with bright sunshine can suddenly be transformed by a blizzard which can last for a week. A blizzard is where snow is picked up by high winds and blown along the surface, severely restricting visibility. Objects less than a metre away may be completely invisible to the naked eye. In these conditions, explorers have died within reach of safety when their shelters have been obscured by the blizzard conditions.

Another hazardous weather condition found in Antarctica is the whiteout. Whiteouts are caused by sunlight reflecting and refracting off snow, ice and water, and they produce disorientating conditions where there are no shadows or contrasts between features in the landscape. The sky and the icy landscape become as one. These conditions are extremely dangerous as they cause a loss of **depth perception** for both humans and birds. Pilots flying in Antarctica have been known to crash their aircraft into the ground, as whiteouts have caused them to feel as though they were flying safely into the open sky. During a whiteout it is almost impossible to distinguish between up and down or horizontal and vertical.

> **KEYWORDS**
>
> Depth perception: the ability to distinguish between objects and distances.

WILDLIFE

The extreme cold of Antarctica and its remoteness from any other land make it a testing and demanding environment for any creature to survive in. For land animals, the surface of the continent is an unremittingly hostile place. However, some enterprising wildlife, both on the land and in the sea, have adapted to life in this freezing world.

All penguin species in the wild are found only in the **Southern Hemisphere**, with seven species of penguins found in the Antarctic. These include the large Emperor and King penguins, as well as the smaller Adelie and gentoo penguins. The greatest numbers of penguins are found on the Antarctic coastline and **subantarctic islands**,

> **KEY FACTS**
>
> Southern Hemisphere: the half of the Earth found below the Equator.
>
> Subantarctic Islands: the small islands found around Antarctica.

such as South Georgia. As a flightless bird, the penguin is well adapted to life in the sea, with its waterproof coat of dense feathers and thick layers of fat offering insulation against the cold. The penguin's main sources of food are fish, squid and crustaceans. On land the penguin is often seen as clumsy and ungainly – the explorer Aspley Cherry-Garrard described them as being 'extraordinarily like children, these little people of the Antarctic world' and early Antarctic explorers were often amused by their comic gait. However, naturalists describe penguins as a miracle of design, with their streamlined bodies making diving for food as easy as flying underwater. The penguin's distinctive black and white colouring helps to camouflage and disguise them from predators such as the leopard seal and the killer whale.

Figure 1.2 The male Emperor penguin incubates the egg during the freezing Antarctic winter

Of all the penguin species, the Emperor penguin is unique as it is the only bird that breeds during the harsh Antarctic winter. When the female Emperor penguin lays her egg she passes it to her mate who incubates the egg on his feet for more than 60 days. The entire penguin colony huddle together for warmth and protection on the ice, while the icy winds whip around them and the temperature continues to drop in the total darkness of the Antarctic winter. If the male were to leave the egg at any stage it would freeze in seconds.

The seas of Antarctica are also home to many species of seals and whales which form an important part of the Antarctic food chain. Whales in Antarctica belong to two groups, toothed whales and baleen whales. Toothed whales, like the sperm whale, feed on fish and squid, whilst the predatory killer whale preys on fish, penguins and seals, often crashing through the floating pack ice to catch unwary seals and penguins.

Although marine animals, seals leave the water for the ice to breed and rest. Of all the seal species, the Weddell seal is found furthest south in the Antarctic waters, breeding on the sea ice in the Antarcic spring. Weddell seals feed mainly on fish, squid and krill, whereas the larger leopard seal, weighing in at 300 kg and measuring up to three metres in length, is a fearsome predator that feeds on penguins and other seals.

THE DISCOVERY OF ANTARCTICA

The existence of the Antarctic continent has been speculated upon for thousands of years. The ancient Greeks believed that the Earth was round, and the Greek thinker Aristotle speculated that the Earth's northern region should be balanced by a similar southern region. He called this region Antarktos meaning 'opposite Arktos', as Arktos was the major star constellation in the northern hemisphere.

Other prominent thinkers in the ancient world agreed with Aristotle's theory. In 100 AD the Egyptian, Ptolemy, wrote a book summarizing the astronomical and geographical knowledge of the ancients and drew a map of the world that showed a large southern continent between the known continents of Africa and Asia. However, these theories were not proven as the idea of travelling to this unknown continent was a terrifying prospect. The spreading belief that the earth was flat, and the support given to this idea by the Church, also prevented further investigation into the existence of a southern continent.

It was not until the sixteenth century and the voyages of the British explorer Captain James Cook that the secrets of the Antarctic regions began to be uncovered. In 1772, under the orders of King George III, Cook set sail in command of two ships, the *Resolution* and the *Adventure*. His instructions were to find the southern polar continent or disprove its existence completely. On his previous voyage, Cook had discovered New Zealand and Australia, now he was to push further south into the unknown.

The *Resolution* and the *Adventure* crossed the Antarctic circle on 17 January, 1773. Cook and his crews became the first people to accomplish such a feat. The seas they encountered were filled with danger for the small wooden ships that Cook commanded. Giant icebergs and floating **pack-ice** dogged the ships' journey, with storms and snow battering the sailors. Cook recorded his wonder and fear at the 'strange mountains of ice' that they sailed through, and became the first but not the last explorer to bemoan the region's temperature '[with] the cold so intense as to hardly be endured …'

> **KEYWORD**
>
> **Pack-ice:** large floating ice-floes that get pushed together.

In the course of his voyage in the Antarctic circle, Cook discovered the island of South Georgia and the South Sandwich Islands. However, the object of his mission remained out of reach, with thick pack-ice forcing his ships back without a sighting of the Antarctic continent. He returned to England disappointed but of the firm belief that the stormy and ice-bound seas that had barred his progress would prevent any other explorers from following in his footsteps, and he confidently stated 'the lands which may lie to the South will never be explored.' The bleak and inhospitable picture of the Antarctic seas painted by Cook dissuaded other adventurers from seeking the fabled southern landmass.

The first sighting of Antarctica fell to a captain in the Russian Navy, Fabian von Bellinghausen. In 1819 Bellinghausen sailed for the Southern Ocean and into the Antarctic circle. Unlike Cook who was thwarted by impenetrable ice and poor weather conditions, Bellinghausen sighted 'an icefield covered with small hillocks' on 27 January 1820. It was the first sighting of the fabled southern continent. The significance of this discovery escaped Bellinghausen, who merely noted the position in the ship's log before continuing with his voyage. Over the course of his explorations in the Antarctic circle Bellinghausen discovered and named the islands Peter I Oy and Alexander Island. Buoyed by his momentous discoveries Bellinghausen returned to Russia where his tales of the cold and foreboding lands he had sighted were met with indifference. Only now in the sea that bears his name is Bellinghausen's contribution to the discovery of Antarctica commemorated.

After Bellinghausen's discovery more and more attention was focussed on the exploration of the Southern Ocean and forays into the Antarctic circle. Merchant seamen, commercial sailors hunting for seals and whales, and navy sailors from Great Britain, the United States, France and other countries all ventured forth in the name of science, commerce and discovery.

THE VOYAGES OF ROSS

The man who added a remarkable chain of discoveries to the body of knowledge about Antarctica was the Royal Navy commander James Clark Ross. An experienced sailor, Ross had great experience of conditions in the polar regions, as in 1831 he was second-in-command on the voyage that discovered the **North Magnetic Pole**. In 1839 the Admiralty gave Ross command of two warships with instructions to sail for Antarctica, with the aim of locating the **South Magnetic Pole**.

> **KEY FACT**
>
> The magnetic poles are created by the Earth's magnetic field and move around slowly due to the effects of the Earth's orbit.

Ross's experiences on his Arctic voyages ensured that the expedition was adequately prepared for the challenge that lay ahead. The warships *Erebus* and *Terror* were specially strengthened, whilst the interior of the ships were refitted to ensure the comfort of the crew during the hard voyage that lay ahead. Ross sailed into the Antarctic circle on 1 January 1841, and soon encountered the thick pack-ice that was the bane of all Antarctic explorers. With his specially reinforced ships, Ross gave the order not to turn back but to press on and try to force a path through the ice-field that stretched before them. Such a daring attempt to penetrate the pack-ice that protected the Antarctic continent had never been tried before and Ross recorded that the ships fought against the ice 'sustaining violent shocks, which nothing but ships so strengthened could have withstood'. Finally, on 9 January they broke through the pack-ice and into open water. Ross had discovered the sea that would be named in his honour. As he continued to sail southwards in search of the South Magnetic Pole, Ross sighted land covered with a sweeping snow-covered range of mountains many thousands of metres high. A boat was landed on an islet and the land claimed for Great Britain in the name of Queen Victoria.

However, Ross's most remarkable discoveries were still to come. The *Erebus* and *Terror* sailed south and eastwards continuing in their mission to locate the South Magnetic Pole. From the ships, an island was sighted topped with a towering and brooding mountainous volcano. The volcano's peak was 3,780 metres high and a column of smoke and flame towered into the Antarctic sky. The doctor on board the *Erebus*, Robert McCormick, recorded the reactions of the crew, 'All the coast one mass of dazzling beautiful peaks of snow ... then to see the dark cloud of smoke, tinged with flame, rising from the volcano. This was a sight so surpassing everything that can be imagined that it caused a feeling of awe to steal over us.' The volcano was named **Mount Erebus** and a smaller extinct volcano to the east of it, Mount Terror.

> **KEY FACT**
>
> Mount Erebus is the only active volcano in the whole of Antarctica.

As the crews continued eastwards an even more awe-inspiring sight greeted them. A sheer cliff of glistening ice, 60 metres above sea level, perfectly level at the top and without any blemishes on its smooth face, stretching as far as the eye could see. In tribute to his Queen, Ross named it the Victoria Barrier, although this name was later changed to the **Ross Ice Shelf**. Although Ross continued his voyage in the hope of finding a route through the 'barrier', he noted in his log that 'we might with equal chance of success try to sail through the cliffs of Dover, as to penetrate such a mass.' After braving fierce storms and terrifying icebergs, the *Erebus* and the *Terror* returned to safe harbour. The South Magnetic Pole, but the importance of his Antarctic discoveries meant that Ross was officially recognized with a knighthood.

From the end of Ross's Antarctic voyages in 1843 until the last decade of the nineteenth century, Antarctica was left to the whaling and sealing industries, centred around the **South Shetland Islands**. However, in 1895 the Sixth International Geographical Congress, a gathering of prominent scientists and representatives of national governments, passed a resolution on Antarctica calling for 'further explorations … before the close of the century'. The race to the pole had begun.

KEY FACTS

Ross Ice Shelf: previously known as the Barrier, it is up to 1000 metres thick.

South Shetland Islands: a group of islands including Elephant Island situated at the northern end of the Antarctic Peninsula.

---

* * *SUMMARY* * *

- Antarctica is the fifth largest continent and one of the coldest and windiest places in the world.

- Antarctic weather is extremely unpredictable with blizzards and whiteouts creating disorientating conditions.

- Wildlife such as penguins and seals visit Antarctica.

- Although the ancient Greeks believed in the existence of Antarctica, it was not until the sixteenth-century voyages of Cook that the seas around Antarctica were explored.

- The Russian, Bellinghausen, was the first person to sight the Antarctic continent in 1820 and the Royal Navy commander Ross added to his discoveries in 1841.

2 In Search of Adventure

Ernest Henry Shackleton was born on 15 February, 1874 in the village of Kilkea in County Kildare, Ireland. He was the second child and eldest son of Henry and Henrietta Shackleton, whose family was eventually to consist of ten children. The Shackleton family originally came from Yorkshire, England, but early in the eighteenth century Abraham Shackleton moved to Ireland and started a school. His son, Henry Shackleton, was at first a farmer in County Kildare and provided his family with a comfortable life. However, Ernest's birth coincided with a failure of the potato crop and so, with agricultural ruin looming, Henry Shackleton uprooted his family from their farm and moved to Dublin to study medicine. Once he had gained his qualifications as a doctor, Henry Shackleton returned to the country of his ancestors, England, bringing his family with him.

The Shackletons settled in London and Henry established a successful medical practice in Sydenham. His wife, Henrietta, however, mysteriously sickened and became an invalid. With work at his practice often taking up much of his time, Henry's mother-in-law and other female relatives assisted with the upbringing of the children. Ernest was educated at home by a governess until he was 11 years old, after which he attended Fir Lodge Preparatory School, where his distinctive Irish accent marked him out as an outsider. On one St Patrick's day, Ernest and another Irish pupil were forced to have a fight. This instilled in him a firm awareness of his Irish ancestry, as well as cultivating a strong sense of justice.

After Fir Lodge, Ernest attended Dulwich College. This public school educated the middle-class children of the neighbourhood and was well suited to producing businessmen, civil servants and writers. However, in later life Shackleton recalled that 'I never learned much' and complained of the constricting nature of the teaching that made even the enjoyment of poetry 'a task and an imposition'. At Dulwich, he surrounded himself with a small group of friends and although his love of practical jokes and occasional truancy often found him in trouble, his good humoured and quick-witted nature always helped him to escape it. Uninspired by his schooling, Shackleton retreated into books and magazines and was inspired by stories of adventure and daring such as *Twenty Thousand Leagues under the Sea* by Jules Verne. This world of imagination was, for Shackleton, an indication of the path he should follow in the real world. Although his father longed for him to become a doctor, Shackleton had decided that he would go to sea.

LEARNING THE ROPES

At this time, there were two ways in which a boy could pursue a career at sea, either by joining the Royal Navy or following a path into the Merchant Navy (commercial shipping fleet). The finances of the Shackleton family prevented Dr Shackleton from sending his son to the Naval Training Academy at Dartmouth. However, with the assistance of a cousin, his father was able to find Shackleton a post on a fully rigged sailing ship, the *Houghton Tower*. At the age of 16, Shackleton joined the crew of the *Houghton Tower* in Liverpool and set sail on a voyage to Valparaiso in Chile. Shackleton's maiden voyage took him around the feared **Cape Horn**.

> **KEY FACT**
>
> Cape Horn is the southernmost tip of South America, surrounded by some of the stormiest and most dangerous seas in the world.

In a letter home to a school friend Shackleton confided that at sea 'you carry your life in your hand.' The sense of isolation onboard the ship was heightened by the fact that at this time there was no radio to maintain contact with the outside world. After calling at Valparaiso and Iquique in Chile for further cargo, the *Houghton Tower* returned to Liverpool, England in April 1891, with food and water running low. It was a true introduction for Shackleton to the hardships of life under sail at sea. In later years Shackleton confided that the seeds of his desire to become an explorer had been planted on that first voyage.

After this voyage, Shackleton signed the **indenture papers** which committed him to train as a Merchant Navy officer. He was to spend the next five years of his life out on the open seas. His subsequent voyages took him to America, South America and the Far East, as Shackleton progressed from apprentice to Second and then First **Mate**. However, Shackleton's elevation to an officer did not stop him from forging friendships with sailors of all positions and classes, even though officers were not meant to mix with engineers or other 'social inferiors'. At sea, Shackleton became adept at handling all types of characters and extremes of behaviour, as he related in a letter to a friend, 'Only the other day I saw a man stab another with a knife in the thigh right up to the handle.' Although young, Shackleton managed to avoid some of the seamier temptations of life at sea – alcohol and women – through his quiet study of books and attempts at writing poetry. In April 1898, at the age of 24, Shackleton was certified as a Master, and qualified to command a British ship anywhere in the world.

> **KEYWORDS**
>
> Indenture papers: a contract committing a person to train as an apprentice.
>
> Mate: Officer on a merchant ship.

Although Shackleton's career at sea was coming into bloom, he craved fulfilment in other areas of his life. Through a visit to his sister Kathleen in 1897 he had fallen in love with Emily Dorman, a friend of his sister, some six years his senior. A long Victorian courtship followed, carried out over several years through letters and visits to the Dormans' family home. In an effort to improve his financial circumstances and make himself more attractive as a marriage prospect, Shackleton joined the Union Castle Line, the prestige fleet of the Merchant Navy service.

His service with the Union Castle Line brought Shackleton many benefits, as their ships ran to a regular timetable as they transported mail between Britain and the outposts of Empire. This meant that he was home every two months and had many opportunities to romance Emily. A fellow shipmate recorded his impressions of Shackleton at this time as being a man of many dimensions, '… he never stood aloof … he was very human, very sensitive,' but also noted that at times Shackleton gave an impression of steely intensity '... another Shackleton, with his broad shoulders hunched – his square jaw set – his eyes cold and piercing; at such a time he might have been likened to a bull at bay.' Shackleton's bullish determination and the respect that he had for his fellow sailors of all classes was illustrated on one voyage when he spoke out on behalf of the cook who had rescued a drowning man. The officer accompanying the cook on his rescue mission was awarded a life-saving medal, whilst the cook was ignored by the authorities. Shackleton was one of the few men who spoke on the cook's behalf. This incident illustrated how Shackleton's nature encouraged him to speak out against injustice and stand up for brave men whatever their background.

SCOTT AND THE *DISCOVERY* EXPEDITION

Although Shackleton was winning the respect of those he served with in the Union Castle Line, he still did not have sufficient means to convince Emily's father, Charles Dorman, a wealthy solicitor, that he was worthy of her hand in marriage. In a bid to dramatically improve his fortunes and fulfil his childhood ambitions of adventure, Shackleton wrote to volunteer to join the National Antarctic Expedition, organised by Sir Clements Markham, KCB, the President of the **Royal Geographical Society**.

As a former Royal Navy officer, who had served on Arctic missions, Sir Clements's mission was to revive the tradition of Naval exploration of Antarctica, which had subsided after Ross's expeditions earlier in the century. For Sir Clements, polar exploration embodied the glory of hard labour and self sacrifice. However, political tensions meant that the Royal Navy was unwilling to commit to a costly and dangerous mission of Antarctic exploration. To realize his dream Sir Clements had to attract sponsors for the expedition to raise the finances required and, by 1899, after financial help from prominent businessmen and the British Government, the National Antarctic Expedition was announced. Unfortunately the dream of a Navy-led expedition was still beyond Sir

> **KEY FACT**
> The Royal Geographical Society: a geographical institution committed to British exploration.

Clements's grasp, until the afternoon of 5 June 1899 when an ambitious 32-year-old Naval officer, Robert Scott, arrived at Sir Clements's home and volunteered to lead the Antarctic expedition.

Scott was a highly ambitious Lieutenant who was eager for advancement in the Navy, to improve his social and financial standing. He saw the prospect of leading an Antarctic expedition as a guaranteed route to promotion. With Sir Clements speaking on his behalf, the Royal Geographical Society and the **Royal Society**, joint co-ordinators of the expedition, appointed Scott to the post of the Commander of the National Antarctic Expedition on 25 May 1900.

> **KEY FACT**
>
> The Royal Society: a scientific academy interested in scientific research.

With £90,000 raised for the Antarctic mission, work commenced on the building of the expedition ship, the *Discovery*, as well as the recruitment and training of the expedition members. Scott tried to ensure that men he saw as loyal to him were selected for the expedition, such as the expedition's junior surgeon Edward Wilson, even overruling a medical report that showed that Wilson had tuberculosis scars on his lungs. Privately Scott expressed 'grave doubts as to my own ability to deal with any other class of men.' However, the Royal Navy, fearful of becoming embroiled in a disastrous mission, granted permission only for three other Naval officers – Charles Royds, Michael Barne and Reginald Skelton – and 20 petty officers and ratings. The rest of the crew was made up of Merchant Seamen.

In February 1901, Shackleton was appointed as a member of the National Antarctic Expedition, and in March, after completing his service with the Union Castle Line, he arrived at the expedition offices to assist with the preparations for the expedition. Shackleton and Scott were similar in a number of ways. Both were fiercely ambitious, seeing the Antarctic expedition as a route to fame and fortune. Also, whilst Shackleton at 26 was six years Scott's junior, in terms of seamanship they were equals, with Shackleton's globetrotting voyages perhaps giving him the edge over Scott's experiences of naval missions and exercises.

Shackleton's place on the expedition was assured as he busied himself in the expedition office and his experience with sailing ships meant that he was put in charge of the *Discovery*'s **trials**. Sir Clements himself was extremely impressed with

> **KEYWORD**
> Trials: test runs to test the ship's performance.

Shackleton's zeal and intelligence, noting in his diary, 'He is a steady high-principled young man full of zeal, strong and hard working and exceedingly good tempered ... A marvel of intelligent energy.' The qualities Sir Clements described would be well suited to an expedition leader, but he also remarked that Shackleton was 'remarkably well informed considering the rough life he has led', giving an insight into the possible reasons why Shackleton was not considered for the command of the expedition.

Previous Arctic exploration had shown the strengths of the use of dogs and skis in polar conditions. Sir Clements, however, saw the use of dogs as cruel and ill-fitted to his ideal of glorious human struggle. The famous Arctic explorer and pioneer Fridtof Nansen argued that '... with dogs it is easier ... it is cruel to take dogs; but it is also cruel to overload a human being.' Scott, after seeking advice from Nansen, decided to take some dogs and skis on the Antarctic expedition. However, few training lessons were given to the crew of the expedition before the start of the mission. Finally, on 6 August 1901, the ship was inspected by the new king, Edward VII, and, with all the preparations complete, the *Discovery* set sail from Cowes, on the Isle of Wight, bound for the Antarctic. She was sailing into troubled waters.

THE VOYAGE TO ANTARCTICA

On the voyage to Antarctica, Frank Wild, one of the crew members noted that it was a journey that 'was neither eventful nor happy.' The ship sprang a leak and her speed was slow and inefficient. Shackleton, a man experienced in the ways of sailing ships recorded in his diary, 'The ship sailed badly ... I am afraid this ... will be a serious matter when we get down South.' Shackleton had to oversee the re-stowing of the supplies and equipment in the leaking hold. This was backbreaking work, and didn't bode well for the rest of the voyage. Only after the brief stopover in New Zealand was the ship able to be fully patched-up and repaired.

Figure 2.1 Explorers had to wear protective clothing to prevent frostbite

The beginning of the final part of the voyage to Antarctica was marked by tragedy. As the ship left Lyttelton harbour, a drunken sailor fell to his death from the main mast. Problems with drunkenness and ill-disciplined behaviour continued to be evident on board, even after such an accident. The *Discovery* was organized like a Royal Navy ship, with officers clearly removed from the ordinary seamen. Shackleton, as a Merchant Navy man, was unused to such divisions.

On 3 January 1902, the *Discovery* crossed the **Antarctic Circle** and headed into the pack-ice. On board the ship, there was tension as the crew wondered nervously about the plans for the expedition. Scott kept the crew in the dark and revealed in a goodbye letter to the explorer Nansen '… I am distinctly conscious of want of plan. I have a few nebulous ideas centring round the main object, to push from the known to the unknown.' Such lack of foresight and planning could be fatal in the cruel Antarctic regions.

> **KEY FACT**
> Antarctic Circle: an imaginary line drawn around Antarctica and its seas.

As the ship forced its way through the belt of pack-ice, the crew had several opportunities to practise their skiing skills on the ice alongside the ship. Penguins and seals on the ice floes were caught and killed for food and added to the expedition store of provisions.

On 30 January, Shackleton sighted land which was named King Edward VII Land (now known as the King Edward VII Peninsula) by Scott. The sighting of the Antarctic continent and the joy of discovery had a profound effect on Shackleton and he recorded that it was a 'unique sort of feeling to look on land that has never been seen by human eyes before.' The *Discovery* continued eastwards in an attempt to add to the earlier discoveries of Ross, passing under the smoking summit of Mount Erebus. As the *Discovery* traced the mountainous wall of ice of the Ross Ice Shelf late on 1 February, fog descended and the ship became trapped in the swirling pack-ice. As the ship circled in vain looking for a way out of the ice trap it was Shackleton, after many hours of trying, who managed to navigate an escape. Shaken by the experience, the *Discovery* turned westwards again and headed for **McMurdo Sound** where the expedition party was to winter.

> **KEY FACT**
>
> McMurdo Sound: an open water bay on Ross Island connected to the Antarctic mainland by the Ross Ice Shelf.

✳ ✳ ✳ *SUMMARY* ✳ ✳ ✳

- Ernest Shackleton was born in Ireland in 1874, but he was brought up in London, England and at the age of 16 joined the Merchant Navy.

- During his five-year service with the Merchant Navy, Shackleton impressed many of his colleagues with his steely determination and selfless personality.

- In 1901 Shackleton was accepted as a member of the *Discovery* expedition, jointly organized by the Royal Geographical Society and the Royal Society and led by the Royal Navy officer Robert Scott.

- While the expedition ship was exploring the Antarctic coastline it became trapped in the pack-ice. With the assistance of Shackleton's navigational skills the *Discovery* was able to escape and anchor at McMurdo Sound.

The Struggle for Supremacy 3

The *Discovery* entered McMurdo Sound on 8 February 1902, before the onset of **Antarctic winter**. Several of the ship's crew were displeased to discover that they were to stay on the frozen continent, having signed up for the posting under the assumption that they would be returning to New Zealand after depositing the expedition. The ship anchored in a sheltered inlet under the foot of Mount Erebus, and was secured to the shore by steel cables, ready to be frozen in for the winter months.

> **KEY FACT**
>
> Antarctic winter: begins in March and lasts until September. At the height of midwinter in June, there are many days of total darkness.

Work commenced to unload the ship and make the necessary preparations for the coming harsh winter. The huts which had been brought to shelter the shore party were assembled and used to store equipment and provisions, as well as to kennel the dogs. The crew stayed onboard the *Discovery* with the ship serving as their living quarters during the winter. Home for the officers was the ship's wardroom, measuring nine metres by six metres and equipped with a large stove, dining table and a piano to provide entertainment. Scott and his officers could also retire to the privacy of their own cabins, whilst the crew's quarters were above the hold.

This separation between officers and crewmen extended to mealtimes, with both groups eating at different times. Breakfast consisted of porridge with bread, butter and jam. Lunch was soup or tinned meat, with supper being made up of the leftovers from the day's meat dishes. **Seal** was also introduced into the men's diet, often replacing the tinned meat or frozen mutton brought over from New Zealand. For all the men,

> **KEY FACT**
>
> The fresh meat that **seals** provided was an excellent source of vitamin C.

the seal meat was considered a poor substitute for the popular, yet limited, supply of mutton. Even though their meals were the same, apart from small luxuries reserved for the officers, the eating arrangements helped to foster resentment and a sense of hardship on the part of the ordinary crewmen.

As the Antarctic season rolled onwards towards the days of absolute darkness, everyone equipped themselves with skis and took to the slopes surrounding the base. Reginald Koettlitz, the expedition's senior surgeon, was pressed into service as a ski instructor. Many of the men found skiing hard to master and received minor scrapes and injuries on the snowy slopes, with the steward, Charles Ford, breaking his leg. Although he escaped injury, Shackleton was pronounced the worse skier of the party. However his innate desire to succeed in everything that he attempted was revealed in a private note that he made to himself, 'Must practise the more.'

Attempts were made to train the dogs to work effectively as sledge teams, a practice that would be vital in transporting the supplies and equipment needed for the southbound expedition. Albert Armitage, the other Merchant Navy officer, who had experience of **dog-driving** in the Arctic argued that the only way to make the dogs pull the loaded sledges was to whip them. Scott and many of the other men viewed such behaviour as cruel and barbaric and preferred to shoulder the burden themselves by practising **man-hauling** the heavy sledges.

> **KEYWORDS**
>
> Dog-driving: the use of dogs to pull loaded sledges.
>
> Man-hauling: when people drag sledges on foot.

As March began the *Discovery* was beset by a terrible tragedy. A party led by Royds was sent to leave messages for the recovery ship, *Morning*, at Cape Crozier 40 miles away at the other end of Ross Island. On their return to the ship a blizzard had engulfed the men and, in a desperate attempt to make it back to the *Discovery*, the men slipped on a steep slope, with Seaman George Vince falling into the sea. When the survivors returned to the ship with news of the accident, Shackleton set out in a small boat to search for Vince around the base of the deadly cliffs where the fall had occurred. Despite his search, Vince's body was never recovered.

THE SOUTH POLAR TIMES

A glum mood spread over the ship after the tragedy and this wasn't helped by the onset of the Antarctic winter with its 100 days of darkness. Shackleton was instrumental in the efforts made to raise the morale of the crew. With assistance from Wilson, Shackleton produced and edited five issues of an expedition newspaper *The South Polar Times*. A mixture of poetry, jokes and cartoons, as well as scientific articles and features, Shackleton welcomed contributions from all of the men on the *Discovery*, whatever their rank or standing. Scott noted the effort that Shackleton put into the production of *The South Polar Times* when he recorded that as well as being editor Shackleton '… is also printer, manager, type-setter and office-boy.' The newspaper played its part in raising morale with Shackleton noting that the first issue '… was greatly praised.'

However, at times morale on board the *Discovery* was not so harmonious. Scott still ran the ship along strict naval lines and even in the freezing winter months, he forced the crewmen to parade on deck for inspection. This resulted in one crew member having most of his toes frostbitten while waiting for Scott to carry out his inspection and led to an atmosphere of distrust and resentment among the ordinary sailors. A steward on board noted that '… many are short tempered and dispirited.' This pointless routine infuriated Shackleton, as one of the naval **ratings** later described in interview '[Shackleton] hated any sort of formality,' and he could often be found on the mess deck talking with the men there. In the eyes of the men below decks, Shackleton appeared to be a more approachable and reassuring leader than the remote and inflexible Scott.

> **KEYWORD**
> Ratings: non-commissioned sailors.

Scott's mind was preoccupied by the choice of the party to journey southwards in the coming summer. Scott had decided that he would lead the expedition himself but, as May became June, he dithered over his choice of companions. Finally on 12 June, Scott revealed his plans for the southern journey to Wilson and invited him to accompany him on the trek. A steward on board the *Discovery* later described the qualities that drew Scott to Wilson: 'Full of constant thoughtfulness for others ... he was the bravest and most unselfish man I have ever known.' Scott's plan was to use all of the dogs in an attempt to get as far to the South Pole as possible.

Wilson was pleased to have been selected by Scott, but pressed him to choose a third member of the expedition party, to safeguard against unforeseen accidents or sudden illness. The close friendship that Shackleton and Wilson had developed whilst working together on *The South Polar Times* led Wilson to suggest Shackleton. Privately, Wilson had reservations about Shackleton's fitness for the trek ahead as he revealed in a letter home to his wife. 'Shackleton hasn't got the legs for the job wants; he is so keen to go, however, that he will carry it through.' Shackleton responded ecstatically to news of his selection. The prospect of being one of the first men to make giant strides across the Antarctic continent beckoned to him, as did thoughts of the fame and fortune it could bring.

As the Antarctic winter slipped away and the sun gradually began to reappear above the horizon, preparations began in earnest for the spring journey. Shackleton was placed in charge of the dogs and ordered by Scott to learn how to drive them. Underfed on a poor diet of dog biscuits the dogs responded unwillingly to Shackleton's amateurish attempts at driving them. With their tails between their legs, the dogs obstinately crawled across the snow despite Shackleton's persuasive urging. Shackleton in turn fretted over his failures, noting day after day in his diary 'dogs refused to pull'.

THE JOURNEY SOUTH

The southern journey finally began at 10 a.m. on Sunday 2 November 1902, with Scott, Wilson and Shackleton leading a team of 19 dogs harnessed to five sledges loaded with equipment and supplies. A team of 13 men, led by Michael Barne, had been sent out three days earlier to lay supplies for them along their route. Initially, Shackleton and the others made good progress and they soon overhauled the advance party led by Barne. However, the men marched across the snow as when the surface of the snow became sticky and cloying, their progress on skis dramatically slowed.

When the weather was fine Scott pushed them into a punishing routine, marching for hours, whilst the dogs strained to pull the sledges. The dogs, however, were ill-suited to this routine, preferring to work in shorter bursts, and they became more unwilling as the journey progressed. Frequently, the men were forced into the harnesses to haul the supplies the dogs refused to pull. Scott described the landscape that they travelled over as 'a broad white plain,' but on many occasions the party encountered adverse surface conditions called **sastrugi**, where the snow has been carved into huge grooves by the wind.

As well as the frustration caused by their slow progress, there was also tension between Scott and Shackleton. Shackleton was responsible for the preparation of the meagre meals that the men ate and one evening, Shackleton accidentally overturned the **primus**, spilling the **hoosh** and burning a hole in the groundsheet of the tent. Scott erupted with anger and blamed Shackleton for his incompetence and insisted that they turn back for base immediately. Wilson was appalled by Scott's behaviour, understanding that their survival depended on their ability to trust and respect

> **KEY FACTS**
>
> Explorers use lines of latitude and longitude to help them work out where they are when travelling. Lines of latitude are horizontal lines around the Earth and measure distances in degrees north and south of the Equator. The South Pole is located at 90° S. Lines of longitude are vertical lines around the Earth giving distances in degrees east and west from the prime meridian, an imaginary line running from north to south around the Earth.
>
> Sastrugi ridges are as hard as rock, and travelling over them on foot or sledge is slow and frustrating. The surface of the ground only appears flat because fresh snow disguises the ridges. This wet and sticky snow afforded the sledges little traction, slowing their progress dramatically.

> **KEYWORDS**
>
> Primus: a portable cooker that burns oil for fuel.
>
> Hoosh: penguin stew.

one another. He managed to calm Scott and convince him that they must all continue. However, the differences that had been hidden between Scott and Shackleton were brought into sharp focus.

Another incident that highlighted the tension between the two men was later recounted by Armitage, a story that he claimed Wilson had confided to him: 'Wilson and Shackleton were packing the sledge one morning. Suddenly they heard Scott shout to them "Come here you bloody fools." They went to him, and Wilson quietly said, "Were you speaking to me?" "No Billy," said Scott. "Then it must have been me," said Shackleton. He received no answer. Shackleton then said, "Right, you are the worst bloody fool of the lot, and every time that you dare to speak to me like that you will get it back."' Under attack, Shackleton was forced to assert his steely personality and trust in his own moral authority.

> **KEY FACT**
>
> Breakfast was bacon fried with biscuit, accompanied by two cups of tea. Lunch another biscuit, with chocolate, sugar and a bovril. Supper was a boiled up hoosh of pemmican (meal and bacon powder), soup and powdered cheese, washed down with a warm cup of cocoa.

As Shackleton became more isolated from Scott he pushed his attention outward to draw motivation and inspiration from Antarctica itself. On 25 November they crossed the eightieth **parallel** and began to fill in the blank spaces of the map of Antarctica. Shackleton recorded in his diary his delight at 'finding out the secrets of this wonderful place.' He saw, too, the beauty and poetry in their lonely conquest of the Antarctic wastes. 'What a little speck on the snowy wilderness is our camp, all round white save where the shadows fall on the snow mounds, and the sun shining down on it all … it is a unique experience.'

> **KEYWORD**
>
> Parellel: a line of latitude.

Ahead of them, as they trudged southwards on the Barrier, lay the uncharted Western Mountains. The men's joy at this discovery was overshadowed by their growing unease as the dogs began to succumb to the harsh conditions, one by one. The first dog, Snatcher, died on 10

December and, to conserve supplies, was fed to the remaining dogs. Scott refused to participate in this grisly task and it was left to Wilson and Shackleton to carry out the butcher's work.

STARVATION AND SCURVY – A RACE FOR LIFE

Driven by his desire to set an unbeatable southern record, Scott decided that they would make a final frantic dash south, taking only a month's supplies and leaving the dogs to feed on each other for food. This was a risky strategy as they had originally planned only to be away for ten weeks, whereas Scott proposed trying for 12 weeks, stretching their provisions to breaking point. Despite the reservations expressed by Wilson and Shackleton, all three men left a depot containing their excess supplies and pushed on.

At night, alone in their sleeping furs, dreams of food haunted the men. Shackleton noted in his diary, 'My general dream is that five three-cornered tarts are flying past me upstairs, but I never seem able to stop them ... The Captain – lucky man – thinks that he is eating stuff, but the joy only lasts in the dreams for he is just as hungry when he wakes up.' Such imaginings were an escape from the cold reality of their exhausting journey. On 13 December, Shackleton recorded glumly in his diary: 'Worst day ... heaviest snow. dist. less than 2 miles. Dogs not pulling. No good ... struggle on. Even higher loads too much ... tired with hauling.'

Hunger and exhaustion were not the only symptoms afflicting Shackleton, his daily record of their journey revealed other symptoms 'Fingers very sore ... feet tired.' Wilson carried out an examination of them all on Christmas Eve; he told Scott that '... both he and Shackleton had suspicious looking gums.' This, along with swollen legs, was the tell-tale sign of **scurvy** in an advanced state. Without treatment the disease meant death. Scott, though, was determined to reach his goal and insisted that they continue marching. Despite their own anxieties, Wilson and Shackleton

> **KEYWORD**
>
> Scurvy: a disease caused by a lack of vitamin C which can result in swollen limbs, loss of teeth, depression and, in extreme cases, death.

followed Scott further south. On 30 December, they reached 82° 15. They had beaten the record for travelling the furthest south by over 320 kilometres. As the explorers turned north once more, their trek became a race to survive.

The three men had to reach the **depots** they had laid before their supplies gave out. To do this they would have to match the pace that they had set on the outward journey, but the men were now suffering from the reduction in their rations. Hunger weakened them and scurvy continued to tighten its hold on the group. If they used their dogs for food the fresh meat would have alleviated their suffering, but their sentimentality prevented them.

> **KEYWORD**
> Depot: a place where supplies are kept.

Scott's diary entries now revealed a sense of regret at his decision to drive the group on. He bleakly recorded: 'It is ludicrous to think of the ease with which we expected to make our return journey in comparison with the struggle it has become.' The struggle was greatest for Shackleton who, by mid-January, was in an extremely bad way. With good luck they had managed to reach the first depot just as food supplies were dwindling away. After replenishing their provisions, all three set off on the 150-mile dash for home. On 18 January, Shackleton, complaining of pains in his chest, collapsed as they marched. He had been driving himself on remorselessly, but of the three of them, scurvy had affected him the worst. He became dizzy, breathless and repeatedly coughed up blood. Privately, Scott and Wilson feared that he would not make it back to the *Discovery* alive.

INVALIDED HOME

Shackleton was now spared from the chore of man-hauling and forced to ski alongside as Scott and Wilson toiled in the traces. Even as he struggled for breath Shackleton felt that he was letting the other two men down, 'My trouble weighs on my mind for I would like to be doing more than just going along.' However, Shackleton's condition was so severe that at times he was forced to ride on the sledge, acting as a brake, whilst Scott and Wilson pulled. This increased Shackleton's feelings of isolation, as he was excluded from Scott and Wilson's discussions.

Battling on in search of their final depot, the weather worsened on 28 January as gales from the south descended, cutting off the clear path ahead. With food supplies exhausted, it was essential that they found the depot and it was Shackleton, leading the way on ski again, who sighted the black flag that marked its position.

At the depot the group set up camp and as the weather around them worsened, so did Shackleton's health. Unable to move, he lay in his sleeping furs, fighting for breath. The next day Scott and Wilson decided to press on for the ship, with Shackleton attempting to ski and then riding on the sledge as he wheezed and coughed up blood.

On 3 February, under the shadow of Mount Erebus, they were met by two of the *Discovery* crew, Skelton and Bernacchi. They were six miles from the ship. Bernacchi recorded the distressed physical condition of the expedition party: 'Long beards, hair, dirt, swollen lips and peeled complexions, and bloodshot eyes made them almost unrecognisable. They appeared to be very worn and tired and Shackleton seemed very ill indeed.'

They returned to the *Discovery* to be greeted as conquering heroes. However, their achievements in reaching 82° 15 had not matched Scott's earlier grandiose aims, and disappointment hung in the air. Also awaiting Scott on his return was the relief ship, the *Morning*, with orders for the expedition to return home immediately. The *Discovery*, though, was frozen firmly in the iced-up bay and all attempts to free the ship failed. Scott decided that he would spend another year on the continent, giving him ample time to improve on his Antarctic achievements.

Shackleton, however, was not given the opportunity to do the same. Scott ordered him to return to England as an invalid: 'He ought not to risk further hardships in his present state of health.' In his diary Wilson noted the effect that the enforced departure had on Shackleton '[it] upset him a great deal as he was very keen to stop and see the thing through.'

Shackleton's illness gave Scott the opportunity he needed to remove his rival. The challenge to his authority that Shackleton had offered on the southern journey was not to go unpunished.

Shackleton departed on board the *Morning* on 2 March 1902. As the crew remaining on board the *Discovery* cheered Shackleton on his way, he broke down and cried. In his diary he recorded, '… the sun came out in a blaze of glory and bathed the bergs in lights that were more than splendour: It was so familiar my home fading away.' The great white wastes of the Antarctic continent had left an indelible mark on Shackleton's soul and he vowed that he would return soon.

* * * SUMMARY * * *

- The crew of the *Discovery* wintered at McMurdo Sound and commenced training for the expedition ahead. To entertain the crew Shackleton produced an expedition newspaper *The South Polar Times*.

- Scott chose Wilson and Shackleton to accompany him on the southern journey and the party set off on 2 November 1902. As the journey progressed, the dogs refused to pull the sledges forcing the explorers to man-haul their equipment and provisions.

- Scott and Shackleton clashed on the journey, but eventually the team reached a new southern record at 82° 15. On the journey home the three explorers suffered from scurvy and hunger, with Shackleton, at one point, collapsing with a suspected heart complaint.

- On their return to the *Discovery* the men were greeted as heroes. However, Shackleton was invalided home whilst Scott remained in Antarctica.

The Strength to Turn Back

Shackleton returned to England in June 1903 and walked straight into a controversy. The Admiralty were enraged that Scott had remained in the Antarctic and, as the expedition organizers had run out of funds, they were forced to equip two ships to carry out the relief of the *Discovery*. Shackleton was drafted in to assist with the fitting out of the relief ships, the *Morning* and the *Terra Nova*. The exiled man was now organizing the rescue of his former commanding officer, Scott.

However, Shackleton had his own pressing problems at this time. As a successful polar explorer he had gained a small measure of fame, but he needed to find gainful employment quickly and a satisfactory salary in order to go ahead with his marriage to his sweetheart, Emily Dorman. After a brief flirtation with journalism,

> **KEY FACT**
>
> Royal Scottish Geographical Society: the Scottish geographical institution committed to exploration and research.

Shackleton was persuaded by a friend, Hugh Robert Mill, to apply for the post of Secretary of the **Royal Scottish Geographical Society**. On 11 January, Shackleton was appointed to the post which had an annual salary of £200. With his future secured, Shackleton married Emily in April 1904, and the couple moved to their new home in Edinburgh.

Shackleton threw himself into his new role with the same irrepressible energy that he had brought to polar exploration. His charm and skill at public speaking swelled the numbers attending the Society's lectures and membership rose steadily under his stewardship. However, Shackleton's restless nature was ill-suited to the dusty meeting rooms of the RSGS and in January 1905 he resigned. After an attempt to become a Member of Parliament for the Dundee constituency failed, Shackleton's attention was drawn southwards to the white continent once again.

After Scott returned from Antarctica, Shackleton watched from afar as Scott was elevated to heroic status and promoted to the rank of Captain. In October 1905, though, the publication of Scott's account of the Antarctic expedition, *The Voyage of the Discovery*, re-ignited the animosity between the two men. In his account of the southern journey, Scott portrayed Shackleton's breakdown as the reason for the unimpressive southern record achieved. Shackleton was furious with the references to 'poor Shackleton' and the mention of the time when illness forced him to ride on the sledge. 'Our invalid was so exhausted that we thought it wiser he should sit on the sledge, where ... we carried him.' Moreover, the references to the scurvy that had afflicted the whole party were presented in a way that made it seem as if Shackleton was the first to succumb, and that Scott himself 'was the least affected of the party.'

THE *NIMROD* EXPEDITION

Shackleton felt that he was the victim of a huge injustice and that Scott had called into question his prowess as an explorer, whilst hiding his own failings in leadership. This spurred Shackleton into action and he began to lay the plans for his own polar expedition. His wife Emily, who had given birth to their first son Raymond in January 1905, was reluctant to see him abandon his fledgling family for the harsh wilderness of Antarctica. However, Shackleton pressed on with his plans for an expedition to reach the South Pole and soon was garnering support from several prominent figures, including the rich Scottish shipbuilder, William Beardmore. By February 1907, Shackleton had raised £30,000 for the expedition and approached the Royal Geographical Society to ask for its support.

The RGS, however, mysteriously declined to offer support and, when Shackleton approached Wilson, his loyal friend on the *Discovery* expedition, to ask him to act as his second-in-command he, too, refused to become involved. Shackleton soon found out why. The public declaration of the expedition appeared in *The Times* newspaper on 12 February 1907, with the headline boasting 'NEW BRITISH EXPEDITION TO THE SOUTH POLE'. The article outlined Shackleton's plans to use Scott's old base at McMurdo Sound and his intention of bettering the achievements of the *Discovery* expedition. Soon after this public statement of intent, Shackleton received a letter from Scott accusing him of attempting to foil his own secret plans for a return expedition. Furthermore, Scott ordered Shackleton to abandon his plans to winter at McMurdo Sound and leave the entire Ross Sea area for Scott alone to explore. This letter was followed by another from Wilson also urging Shackleton to give up his plans.

Shackleton felt betrayed and his reply to Scott stated that his enforced return from the *Discovery* expedition gave him the right to plan his own Antarctic expedition, especially as the Secretary of the RGS had revealed to Shackleton that Scott's plans for a return expedition were only tentative. The harsh sniping from Scott and constant stream of appeals from Wilson gradually wore away Shackleton's resolve. He agreed to leave the *Discovery* base at McMurdo Sound untouched and turned his attention eastwards along the Antarctic coastline to **King Edward VII Land** or the **Barrier Inlet** as his proposed base. Scott and Wilson had won this concession, but Shackleton hoped that his expedition would bury Scott's Antarctic achievements by claiming the pole in a blaze of glory.

> **KEY FACTS**
>
> **King Edward VII Land:** now known as King Edward VII Peninsula, it is found at the eastern end of the Ross Ice Shelf.
>
> **Barrier Inlet:** a natural inlet in the Ross Ice Shelf discovered by Scott's *Discovery* expedition.

With his new plans in place, Shackleton swiftly recruited his crew. After he was turned down by many of the fellow officers who had accompanied him on the *Discovery* expedition, Shackleton turned his attention further afield. The qualities that Shackleton looked for in his fellow explorers were optimism, patience, endurance and courage, all qualities that Shackleton himself embodied.

One of the few members of the *Discovery* expedition who did agree to accompany Shackleton on this new adventure was Frank Wild. On the *Discovery*, Wild belonged to the lower decks as an ordinary naval rating, but Shackleton had sought out the wiry and unassuming Yorkshireman as a friend. At the chance to return to Antarctica under Shackleton's command Wild readily accepted.

With the expedition due to set off in August 1907 Shackleton had scant time to organize equipment and supplies. Like Scott before him, he consulted the Norwegian explorer Nansen for advice. Nansen counselled him to put his trust in dogs and skis as the most effective means of polar transport. However, Shackleton's experiences on the southern sledging journey associated skis and dogs with sickness and failure. Shackleton's plan was to walk to the South Pole accompanied by horses. The experiences of the Arctic explorer Frederick Jackson in the polar regions encouraged Shackleton to order twelve hardy **Manchurian ponies**.

> **KEYWORD**
> Manchurian ponies: a type of pony found in the mountainous north-eastern area of China.

Shackleton also proposed to try out a new form of transport in Antarctica, as his wealthy backer William Beardmore had provided him with a specially built motor car. In a newspaper interview Shackleton boasted that 'the machine can travel 150 miles in twenty-four hours.' The motor car, though, was still a relatively new invention and Shackleton was putting a lot of faith in an unproven machine that had yet to experience the extreme conditions of the southern continent.

The most important piece of equipment that Shackleton needed to secure was a ship to transport the expedition to Antarctica. Limited finances meant that he had to scout around carefully for a suitable and affordable vessel. For £5000 he managed to buy a sealing ship named *Nimrod*; 41 years old, the *Nimrod* at 300 tons was one of the smallest ships to head for the stormy Southern Ocean that surrounded the Antarctic.

In the midst of the hectic preparations, Shackleton was haunted by financial worries as several of the expedition's backers withdrew their support, and he was pursued by his many creditors for payment. Shackleton battled his way through these setbacks, employing his charm and tough exterior in equal measures to ensure that preparations went ahead unhindered. **Furs**, sledges and provisions were purchased, with Shackleton often making little improvements to the equipment that Scott had taken on the *Discovery* expedition.

> **KEY FACT**
> Furs: following the example of the native Arctic peoples, explorers used layered furs as insulated clothing.

On Sunday 3 August 1907, the *Nimrod* departed for the Antarctic. On board the ship Shackleton set out the mission for his men. 'War in the old days made men. We have not these same stirring times to live in and must look for other outlets for our energy and for the restless spirit that fame alone can satisfy.' He wanted to inspire his men with the same spirit of adventure and hunger for glory that drove him on.

A BROKEN PROMISE

En route for Antarctica, the *Nimrod* harboured at Sydney in Australia and Lyttelton, New Zealand. The force of Shackleton's persuasive powers led both the Australian and New Zealand governments to provide him with funds to assist with the refitting of the *Nimrod* for the expedition's eventual relief. In order to conserve fuel, the *Nimrod* was towed to the edge of the Antarctic pack-ice by a tramp steamer. After navigating the icebergs and pack-ice the *Nimrod* came out into open water on 20 January 1908. Shackleton now ordered the ship to turn eastwards, away from the territory claimed by Scott, to search for the Barrier Inlet.

Shackleton's daring plan was to winter on the ice shelf itself, which he calculated would move him nearly 200 kilometres closer to the pole. However, as they cruised along the edge of the mighty Barrier they could see no trace of the Barrier Inlet. In its place lay a large bay encircled by sea ice and filled with the spouting signals of whales, which Shackleton dubbed the Bay of Whales. Barrier Inlet no longer existed as part of the ice shelf had broken away, carrying it out to sea.

Shackleton was faced with a tortuous decision. He had three choices: to land on what now seemed to him to be a very impermanent ice shelf; to press on eastwards through the dangerous pack-ice in search of a yet undiscovered base on King Edward VII Land; or to turn back for McMurdo Sound, breaking his promise to Scott. The bond of his word was set against Shackleton's fears for the safety of his men. After a desperate attempt to reach King Edward VII Land failed because of the menacing pack-ice, the *Nimrod* headed disconsolately for McMurdo Sound.

Even when they reached McMurdo Sound, the Antarctic environment conspired against Shackleton, as 25 kilometres of solid sea ice prevented the *Nimrod* from reaching the *Discovery*'s anchorage at **Hut Point**. Shackleton was forced to set up base at Cape Royds, a crop of land under the shadow of Mount Erebus.

> **KEY FACT**
>
> Hut Point: a point of high land on Ross Island named after the *Discovery* expeditions's storeroom.

Early February was spent unloading equipment and supplies from the *Nimrod* and setting up the hut that was to be the expedition's winter home. On 22 February, the *Nimrod* sailed again for New Zealand carrying orders to return to McMurdo Sound in a year's time.

THE ANTARCTIC WINTER

The expedition party of 15 men were completely cut off from the rest of the world. They spent the Antarctic winter in the small hut at Cape Royds, in close company around the stove and long table that dominated the main room. This room was where all work, meals and entertainment took place, with Shackleton keeping a watchful eye that no serious divisions sprang up between the men. Apart from the main living area, the hut was subdivided into small bedrooms which the men shared. The ponies were stabled alongside the hut's sheltered side, protected by walls constructed of bales of fodder and cans of food.

As the winter set in, Shackleton proceeded to plan for what would be the expedition's biggest achievement – the journey south to the Pole. His original plan was to use the ten ponies to haul supplies on the southward journey, but two had died on the *Nimrod*'s voyage from New Zealand, and a further four had poisoned themselves by eating sand from the beach at Cape Royds soon after landing. This left only four ponies – Socks, Grisi, Quan and Chinaman. The motor car that Shackleton had confidently envisaged cruising to the Pole proved to be a disaster, with the wheels churning uselessly whenever they encountered loose snow. These setbacks meant that Shackleton had to revise his plans for the southern journey, cutting back on the number of men he intended to take.

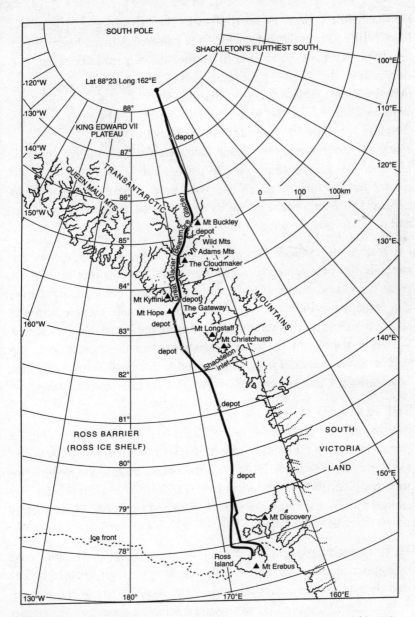

Figure 4.1 The route of the *Nimrod* expedition that took Shackleton to within 160 kilometres of the South Pole

He shared the process of selection with the expedition's doctor, Eric Marshall; Shackleton's experiences on the *Discovery* journey highlighting the importance of good health in the arduous conditions of the Antarctic. Marshall recommended the **meteorologist** Jameson Boyd Adams, Frank Wild and himself for the journey, confident in their ability to take the physical strain of the trek. Marshall had reservations about Shackleton's health, as he was constantly refused permission to examine 'the Boss', as Shackleton was known. There was no question in Shackleton's mind that he was to be denied his dream of polar glory, and he completed the team of four.

> **KEYWORD**
>
> Meteorologist: someone who studies the atmosphere and weather patterns.

Marshall was a valuable member of the expedition, as it was his supervision of the camp's diet, ensuring plentiful consumption of fresh meat, such as penguin and seal, that prevented any outbreaks of scurvy. For Shackleton, this skill alone justified Marshall's inclusion in the southbound party. There was another reason, too – Marshall had been a source of tension in the winter quarters, often disagreeing with the decisions that Shackleton made. By taking Marshall south with him Shackleton was able to keep a close eye on a potential troublemaker.

THE ROAD TO THE POLE

After all the preparations for the southern trek were completed and the depots laid along the southern road, Shackleton's team set out in bright sunshine on 29 October 1908. With the ponies pulling the sledges, the men traipsed on foot, averaging over 24 kilometres a day in the first month out. By the end of November they had travelled 480 kilometres down the road to the Pole.

However, such rapid progress was not without its costs. In the biting cold of Antarctica the ponies suffered, the ice froze on their exposed bodies and their hooves became wounded by the jagged edges of the sastrugi ridges. At night, while the men camped safely in their tent, the ponies were left standing shivering under woollen blankets, virtually unprotected fom the howling Antarctic winds that swept over the ice cap. Shackleton's diary recorded their suffering, '… the ponies struggling gamely … had to plough through truly awful surface.'

On 21 November, the pony Chinaman could go no further as the terrible ice surface had crippled his legs. The ponies Grisi and Quan's strength gave out days later. To spare their suffering the ailing ponies were shot. Their failing was a mixed blessing for the men, as more of the heavy burden of hauling fell upon them, but the meals of fresh pony meat helped to keep scurvy at bay.

By December the party had passed Scott's furthest point south and reached a huge glacier leading through the Transantarctic Mountains. Shackleton named this the **Beardmore Glacier** after his generous expedition backer. The surface underfoot became dangerous as many crevasses

> **KEY FACT**
>
> The Beardmore Glacier is 48 kilometres wide and over 160 kilometres in length.

littered their path. The remaining pony, Socks, was harnessed to a sledge loaded with equipment and supplies as Wild led it up the glacier. Suddenly, a crevasse opened up beneath Socks and the pony plummeted through the ice to its death. Wild fell back into the newly opened crevasse and only his quick reactions saved him as he clung to the edge, before he was swiftly rescued by the others. The sledge, with its cargo of valuable food and fuel, was miraculously saved as the force of Socks's fall had shattered the wooden bar that attached the pony to the sledge.

> **KEYWORD**
>
> Snowblindness: a painful swelling of the eyes caused by the glare of sunlight reflected off ice and snow that results in a temporary loss of sight.

Shackleton and his men could now rely only on their own strength and fortitude to conquer the treacherous Beardmore Glacier. As they scaled the glacier, the increasing altitude and terrible **snowblindness** both took their toll on the group. Every step forward was dogged by the fear of deadly crevasses, as their feet constantly broke through the surface of the ice, with only the harnesses that linked them to the sledge saving them from oblivion.

Christmas Day 1908 saw the men camped at 86° South, 2,850 metres up the glacier. Shackleton allowed them all a celebratory feast consisting of an increased ration of hoosh, topped with a minute portion of plum pudding, a spoonful of brandy and a restorative cigar. Even in times of extreme mental stress and physical strain, Shackleton worked hard to maintain the morale of his men and create an oasis of normality in the alienating wastes of Antarctica.

FURTHEST SOUTH AND FAILURE

As the new year of 1909 dawned, worsening weather ate into the optimism that Shackleton had built up. The team was constantly buffeted by sweeping blizzards as they crossed onto the polar plateau, with the temperature dropping rapidly. Symptoms of severe frostbite developed and, as the rations were reduced to conserve supplies, headaches and nosebleeds became common among the men.

Their increasing hunger reminded Shackleton of Scott's earlier follies on the *Discovery*'s southern sledge journey, pushing on when supplies were low, and his focus turned from the distance remaining to the pole to how far they would have to travel back to reach base camp. Dreams of glory lay ahead on the southward road, but Shackleton's dreams were haunted by a heavy sense of responsibility for the welfare of his men. In his diary, Shackleton bleakly recorded 'We are weakening ... from want of food ... We are not travelling fast enough to make our food spin out and get back to our depot in time.' The dream of reaching the Pole was fading fast.

By 9 January they had reached 88° 23′ South, only 155 kilometres from their goal of the South Pole. However, Shackleton finally admitted defeat confiding in his diary, 'We have shot our bolt.' They planted the Union Jack to mark their achievement and solemnly photographed each other against the flag and an endless backdrop of snow and ice. They were 1168 kilometres from the base at Cape Royds. Shackleton had shown a unique kind of bravery by turning back while there was still a chance of survival.

The return home was a race against time. A southerly wind at their backs hurried them on their way, and Shackleton ingeniously rigged a makeshift sail to their remaining sledge to take advantage of this. The men were forced to run alongside the sledge to keep up, even as **frostbite** and hunger threatened to overwhelm them. With their rations running out, the team hurried on in search of their food depots.

> **KEYWORD**
>
> Frostbite: an injury to the body tissue caused by extreme cold temperature. Severe cases can result in the loss of body parts.

FORTY HOURS WITHOUT FOOD

On 18 January, the party began the deadly descent down the Beardmore Glacier. Their fear now was not of the hidden crevasses, but of failing to reach the depoted food left below them on the glacier and the men raced over the crevasses with reckless abandon. However, Shackleton's great strength was weakening and on 20 January he collapsed. There was no time to stop and let him recuperate, and Shackleton was left to walk alongside as his companions pulled the sledge. The others bore him no resentment, as Wild noted that Shackleton had previously been 'doing far more than his share'.

By 26 January their food was exhausted with the depot still 32 kilometres in the distance. They were trapped on the glacier in a maze of crevasses that slowed their progress to a crawl. Hunger and thirst tormented the emaciated men as they staggered grimly onwards. Shackleton later commented, 'I cannot describe adequately the physical and mental strain.' As 27 January dawned, the game was almost up. Wild and Adams collapsed in their tracks, while Shackleton himself was too weak to continue. They were only 5 kilometres from the depot. Drawing on his last reserves of strength, Marshall went ahead alone to retrieve the life-saving rations for the others.

Replenished, they pressed on and finally reached the Barrier once more. Recovered from the effects of the glacier's high altitude, Shackleton once again took command of the leadership of the group. He led them in a race from depot to depot, as they suffered from crippling attacks of **dysentery**, the result of eating spoiled pony meat.

> **KEYWORD**
>
> Dysentery: a form of blood poisoning resulting in severe diarrhoea.

Even in the midst of such suffering, Shackleton still had the ability to inspire. One evening in the tent, after an arduous march and with the spectre of starvation looming, Shackleton turned to Wild and asked him if he would return to Antarctica with him and attempt to reach the South Pole again. Even with their survival by no means certain, Shackleton was looking ahead to future challenges. Wild, without hesitation, replied 'Yes'. Wild's absolute loyalty to Shackleton had been sealed when, in Wild's worst moments of sickness and hunger, Shackleton forced on him his own breakfast biscuit.

As February wore on, the explorers drew near to their base. Out on the Barrier, Marshall collapsed suffering from painful cramps and diarrhoea and Adams was left behind to nurse him. Calling on all their last reserves of energy, Shackleton and Wild forced themselves forward in search of rescue. After marching for three days, with only a few hours snatched rest, they reached Hut Point and salvation. The crew aboard the waiting *Nimrod* greeted the two men as heroes, but Shackleton's first thoughts were for the two men waiting on the Barrier and he immediately led a team to carry out their rescue.

It was an incredible achievement. Shackleton and his team had reached within 160 kilometres of the Pole, but the ultimate achievement had eluded them. If he had been prepared to sacrifice the lives of his team, Shackleton could have claimed the Pole, but as he later remarked to his wife Emily, 'a live donkey is better than a dead lion'.

Although inwardly Shackleton felt that he had failed, on his return he was still feted as a national hero. Lectures he gave about the expedition were packed, and record crowds attended displays of the *Nimrod* in Liverpool and London. The Government also showed its appreciation of Shackleton's magnificent efforts by clearing the debts that he had amassed whilst organizing the expedition. A final honour elevated

Shackleton above his nemesis Scott, as he was knighted in the King's **birthday honours list**. The name of Sir Ernest Shackleton was acclaimed throughout the world.

> **KEY FACT**
>
> Birthday honours list: a special awards ceremony organized by the British Royal Family.

* * * SUMMARY * * *

- On his return from the *Discovery* expedition, Shackleton became Secretary of the Royal Scottish Geographical Society and attempted to become a member of parliament. Shackleton also planned to lead his own expedition to the South Pole, a plan that was fiercely opposed by Scott.

- Shackleton set out for Antarctica on 3 August 1907. His plan to winter on the Barrier failed as part of the ice shelf had broken away and he was forced to winter at McMurdo Sound, breaking his promise to Scott.

- Shackleton selected Marshall, Wild and Adams to accompany him on the trek to the Pole and the party set out on 29 October 1908. At first the party made good progress but, one by one, the Manchurian ponies hauling the sledges died leaving the men to drag the sledges across the Beardmore Glacier that led through the Transantarctic Mountains.

- On 9 January 1909 Shackleton reached 88° 23' South, 155 kilometres from the South Pole, but was forced to abandon the attempt due to a shortage of food and the effects of frostbite and snowblindness.

- On the return journey Shackleton collapsed but, after recovering, marched for three days non-stop to fetch rescuers from Hut Point. On his return to Britain he was knighted by the King.

5 Endurance

Shackleton spent the years after the *Nimrod* expedition trying to consolidate the fame that he had achieved and to capitalize on his reputation as an explorer with a series of money-making schemes. However, in 1912 news broke that eclipsed Shackleton's Antarctic achievements.

The Norwegian explorer **Roald Amundsen** became the first man to reach the South Pole, beating Shackleton's rival, Scott, to exploration's holy grail. Amundsen had reached the Pole on 14 December 1911, after establishing his base at the Bay of Whales. The discoveries that Shackleton had made on the *Nimrod* expedition aided Amundsen, and he took the risk of establishing his base on the ice shelf that Shackleton had refused. Shackleton was magnanimous in his praise for Amundsen, cabling him with congratulations for his 'magnificent achievement' and writing in the *Daily Mail* newspaper that 'Amundsen is perhaps the greatest Polar explorer.'

> **KEY FACT**
> Roald Amundsen: this accomplished Norwegian explorer led a team of five men and reached the South Pole on 14 December 1911 before returning safely to civilization.

Then, in 1913, the news broke that Scott, Wilson and other expedition members had died on their way back from the Pole. A sudden upsurge of national grief greeted the news. Before it was known that Scott had died, he had merely been viewed as the loser in the race to the pole, but the discovery of his frozen body, and the publication of his expedition diaries turned him into a

> **KEY FACT**
> On Scott's *Terra Nova* Expedition, the British explorer reached his goal of the South Pole only to discover that he had been beaten there by his rival Amundsen. Returning to base, Scott's party struggled through intensely cold weather conditions before becoming trapped by a severe blizzard. Scott's diaries eloquently recorded their struggle before the party died.

vanquished hero. He was posthumously knighted by the King and Shackleton's star began to wane.

PLANS TO CROSS ANTARCTICA

As Shackleton's fame faded, so his financial and personal worries mounted. He had three children and a wife to support, yet still his thoughts returned to the wide open spaces of the Antarctic. Much to the disapproval of his wife Emily, Shackleton began to plan for the ultimate polar expedition – the first crossing of Antarctica. Shackleton named the expedition 'The Imperial Transantarctic Expedition' and the plan was to cross Antarctica from coast to coast, beginning at the Weddell Sea coast; with Shackleton leading his men across the continent via the South Pole to the Ross Sea, a distance of 2,400 kilometres. Shackleton was 39 years old and wrote revealingly in a letter to a friend, 'perhaps the Antarctic will make me young again.'

Legend has it that Shackleton placed an advertisement seeking volunteers for the expedition, the text of which read: 'Men wanted for hazardous journey. Small wages, bitter cold, long months of complete darkness, constant danger, safe return doubtful. Honour and recognition in case of success.' The advertisement was brutally honest about the dangers ahead, but still eager volunteers flocked to the expedition offices in London, desperate to join forces with the inspirational figure of Shackleton.

The first men whom Shackleton appointed to the expedition were three of the crew from the *Nimrod* expedition, Frank Wild, George Marston (the expedition artist), and Thomas MacLeod. Wild kept the promise he had made to Shackleton back in the frozen wastes of Antarctica and Shackleton, so impressed by Wild's loyalty and courage, he made him his second-in-command. As well as his *Nimrod* colleagues, Shackleton found other men with proven polar expertise and experience. He appointed the Irishman Tom Crean, a veteran of both Scott's Antarctic expeditions, as Second Officer and Alf Cheetham, a veteran of three Antarctic expeditions, as **bo'sun**.

> **KEYWORD**
>
> Bo'sun: a ship's officer in charge of equipment and crew.

Shackleton was keen to exploit the financial rewards a successful expedition could bring. The Australian explorer Douglas Mawson had employed the photographer Frank Hurley on his 1911–14 Antarctic expedition. Hurley's striking photographs of the expedition amazed the general public and received great acclaim. Shackleton invited the imposing Australian Hurley to join his transantarctic expedition as official photographer and film-maker and formed a company to sell the rights to Hurley's work. Hurley, a born adventurer, grasped the opportunity to chronicle such a unique endeavour.

> **KEY FACT**
>
> A veteran of Shackleton's *Nimrod* expedition, Mawson headed the Australasian Antarctic Expedition that discovered King George V Land, before enduring a harrowing trek across Antarctica after his supplies were lost down a crevasse.

THE *ENDURANCE*

Shackleton was fortunate in his acquisition of the expedition's ship, as a specially built ship, the *Polaris*, had been ordered for a Norwegian company planning to run tourists to the Arctic. After the company collapsed the *Polaris* languished in a Norwegian shipyard awaiting an owner. The ship seemed ideal to Shackleton, with there even being a darkroom aboard for the photographer Hurley. He purchased the ship and changed her name from the *Polaris* to the *Endurance* in tribute to his family motto 'Fortitude Vincimus – By endurance we conquer.'

> **KEY FACT**
>
> The ship had a 350-horsepower steam engine and could reach a top speed of 10.2 knots. Specially constructed to withstand the pressure of the pack-ice, the *Polaris*'s keel was over two metres thick and made out of solid oak, whilst her sides were made out of oak and fir protected by greenheart, a toughened wood heavier than iron.

The important post of ship's captain was filled by Frank Worsley, who later remarked that he had been inspired to join the expedition because of a dream where he was navigating a ship down an ice-filled Burlington Street in London, where Shackleton's expedition offices were located. An impulsive and unpredictable New Zealander, Worsley was a proven sailor and adept navigator.

Shackleton's preparations for the expedition were overshadowed by ominous world events. The increasing political tension in Europe meant that Shackleton was able to secure only one Royal Navy officer for the expedition, Captain Thomas Orde-Lees of the Royal Marines, an accomplished skier and climber. On Monday 29 June 1914, the heir to the Austrian throne, Archduke Franz Ferdinand, was assassinated, setting in motion the chain of events that led to the outbreak of the First World War. As nation after nation was drawn into the conflict, the headlines of British newspapers filled with the unfolding drama.

On Saturday 1 August 1914, as the *Endurance* sailed from London on the first leg of her journey, Germany declared war on France. It was only a matter of time before Britain would be forced to enter the fray. Shackleton was torn between his desire to press ahead with his Antarctic adventure and his patriotic longing to serve his country. With the full agreement of the crew he cabled the Admiralty offering the full services of the *Endurance* and his men to the country in the event of war. Winston Churchill, the First Minister of the Admiralty, thanked Shackleton for his offer, but instructed the *Endurance* to proceed with the expedition, seeing it as a way of boosting the morale of the nation at a time of crisis. The *Endurance* then sailed for Buenos Aries, the first staging post on her voyage to the Antarctic.

After arriving at Buenos Aries, four unruly crewmen were dismissed and, after final preparations for the expedition were made, the ship headed for South Georgia, her last port of call before Antarctica. Three days into the journey to South Georgia, Shackleton discovered a stowaway, Perce Blackborrow, a 19-year-old Welshman, who had hidden himself on the ship at Buenos Aries. Shackleton gave Blackbarrow a fierce dressing-down, accusing him of putting the safety of the entire expedition at risk, before menacingly adding that if the ship's crew became hungry a stowaway was always the first to be eaten. Blackborrow, unbowed by Shackleton's tirade, wittily responded 'They'd get a lot more meat of you, sir.' Impressed by the lad's spirit and amused by his quick humour Shackleton signed him up for the expedition.

INTO THE WEDDELL SEA

At South Georgia, bad news greeted Shackleton. The *Endurance* harboured at Grytviken, a small whaling station staffed mainly by Norwegian sailors. These experienced whaling men informed Shackleton that the Antarctic pack-ice was unusually bad in the Weddell Sea that year. Shackleton heeded the warnings and waited until 5 December 1914, with the Antarctic summer well under way, before embarking on the fateful voyage to the white continent.

Three days into the voyage to Antarctica, Shackleton sighted the first pack-ice, much further north than he had expected to encounter it. On 11 December, the ship's captain, Worsley, was forced to enter the pack-ice and began to navigate the *Endurance* through narrow **leads**. As the ship moved southwards, at a rate of 48 kilometres a day, the pack-ice conditions worsened and Shackleton recorded 'Very large floes ... presented a square mile of unbroken surface ... Steering under these conditions required muscle as well as nerve.' In Worsley, Shackleton had a nerveless navigator who at times would force the *Endurance* through the pack by ramming the **ice-floes**. Whenever the going was especially dangerous and fraught, Shackleton ensured that he was present on the bridge to ensure that no unnecessary risks were taken.

Heavy pack-ice continued to slow the progress of the ship through late December, although Shackleton made sure that the crew's spirits remained high on Christmas Day by celebrating with a dinner for everyone. Music, drinks and a sing-along meant that a mood of togetherness and good humour prevailed. However, among the crew the first notes of concern

> **KEY FACT**
>
> Earlier in the century two polar explorers, Nordenskjold and Filchner, had both seen their ships trapped by the treacherous pack-ice in the Weddell Sea. Conditions in the Weddell Sea, situated as it is between three stretches of land – the Antarctic continent to the south, the South Sandwich Islands to the east, and the Palmer Peninsula to the west – mean that ice forms in any season and is swirled in a clockwise direction by the prevailing current and south-easterly winds. Any ships trapped by the Weddell Sea ice risk being crushed by the pressure of the pack-ice.

> **KEYWORDS**
>
> Leads: channels of open water among the pack-ice.
>
> Ice-floe: a large sheet of floating ice.

were sounded about the *Endurance*'s slow and tricky progress through the pack-ice, with Lionel Greenstreet, the Chief Officer on the ship, writing ominously in his diary, 'Here endeth another Christmas Day, I wonder how and under what circumstances our next one will be spent.'

As the year 1915 began, the *Endurance* was forced to backtrack, following leads in the ice northwards and westwards, whilst Shackleton privately fretted over the lack of progress southwards towards land. On 6 January, the ship anchored next to a large ice floe and the dogs were taken out to exercise on the floe, while some of the crewmen had an impromptu game of football on the ice. At one point Worsley had to be fished out of the sea after falling through a patch of rotten ice.

By 10 January, as the *Endurance* passed through the loose ice of the pack, land was sighted to the south-east. This was **Coats Land** and the ship began to track its coastline southwards along a high wall of barrier ice. A sense of excitement and expectation gripped the crew as they calculated that they were a little under a week away from their destination at Vahsel Bay.

> **KEY FACT**
>
> Coats Land: land to the east of the Ross Ice Shelf, discovered by the Scotsman William Bruce.

Late on 15 January, Shackleton made a decision that he was later to regret bitterly. The *Endurance* had found a sheltered bay protected from all but northerly winds, that would have made an excellent landing place, but Shackleton decided to press on south for Vahsel Bay. As grounded icebergs along the coastline barred the *Endurance*'s path, the ship was forced to turn away from the coast to go around them. The ship's progress was impeded, first by a blizzard that blew up on 17 January and then by the changing conditions of the pack-ice. Worsley noted that, 'The character of the pack has again changed … The floes are very thick [and] the **brash** between is so thick and heavy that we cannot push thro'.'

> **KEY FACT**
>
> Brash: a type of ice formed from the wreckage of larger pieces of ice.

The gale-force winds had driven the pack-ice around the ship and Shackleton recorded glumly on 20 January that, 'As far as the eye could reach from the masthead the ice was packed heavily and firmly all round the ship in every direction.' As January wore on, a crack appeared in the ice in front of the ship, less than 90 metres away, but even with the engines at full steam ahead and the sails raised, the *Endurance* remained frozen solid in the ice. The ship's crew even climbed out onto the pack-ice with chisels and saws and attempted to cut through the ice to free the ship.

On 27 January, Shackleton gave the order to put the engine room fires out, in order to preserve the remaining stocks of coal. As the temperatures dropped, cementing the ice-floes back together, Hurley wrote, 'It appears as though we have stuck fast for this season'. The Antarctic summer was ending and with it all hopes of an early escape from the ice.

Frozen fast, the *Endurance* drifted along with the currents of the Weddell Sea and at the end of February reached the southernmost point of her drift, a tantalizing 40 kilometres from Vahsel Bay, the expedition's intended base. On board, Shackleton accepted that the ship was trapped for the winter, but worried about where the current would carry them and whether, when the spring finally arrived, the ship would be able to free itself from the pack-ice and find land. Macklin, one of the ship's surgeons, noted in his diary that, 'Shackleton at this time showed one of his sparks of true greatness. He did not rage at all, or show outwardly the slightest sign of disappointment; he told us simply and calmly that we must winter in the Pack, explained its dangers and possibilities; never lost his optimism, and prepared for Winter.'

TRAPPED

On 24 February, the ordinary routine of the ship was halted and the *Endurance* became a **winter station**. The stores were diligently checked so that Shackleton was aware of the provisions he had to hand for the coming winter, and the dogs' kennels were moved off the ship onto the floe to allow the dogs space to exercise. In the isolation of their surroundings,

> **KEYWORD**
> Winter station: a base shut down for the winter.

the crew's enemy was boredom and despair. To combat this Shackleton established a strict winter routine, knowing that to improve their chances of escape and survival, the men needed to be kept active and alert. Their ultimate goal now was not the crossing of the Antarctic continent, but the earliest possible escape from the ice. All hands worked by day and slept by night, save for a solitary night watchman, who was instructed by Shackleton to report any changes in ice conditions that could mean an opportunity to free the ship.

To maintain morale among the men, Shackleton issued the winter clothing originally reserved for the expedition shore parties. The warm wool shirts, jerseys, underwear and mitts, along with the windproof Burberry tunics and trousers, helped to protect the crewmen from the icy conditions, and, more importantly, showed the concern that Shackleton had for all crew members. This gesture ensured that Shackleton had the respect of the crew, and he fostered this trust by encouraging the men to approach him over any matter, no matter how small. In this way, Shackleton kept both the friendship of his men, but was also able to monitor closely their moods and feelings.

The ship's carpenter, Harry McNeish, was ordered to build winter quarters for the crewmen on the deck of the *Endurance*. These two by one-and-a-half metre cubicles housed two men each, and were insulated from the extreme cold found in other parts of the ship. Shackleton's quarters, in the Captain's cabin, were situated in the coldest part of the ship, but he accepted his situation without complaint, thus setting a stoic example to the other men.

All through March the *Endurance* continued to drift northwards and as the nights began to lengthen with the onset of the winter, Shackleton organized the men and dogs into six teams. Many hours were now spent out on the pack-ice training the 50-odd dogs. The companionship that the dogs provided for the men was vital in the lonely wastes of the pack-ice. Missing their families and loved ones, the large and playful dogs became surrogate children to the crewmen who cared for them. The races between the dog teams were also a source of entertainment to the men, with many showing a great sense of pride in their own dogs achievements. Hurley revealed in his diary, 'My team is one of the best.'

With the disappearance of the sun in May, the men were forced to seek different forms of entertainment on board the *Endurance*. Sing-songs, reading, games of chess and debates all helped to keep up spirits. Orde Lees recorded in his diary, 'We all manage to live very happily here on board in spite of ... the fact that most members are what one might term very distinct personalities.' The melding of these distinct personalities into a harmonious team during the most trying of circumstances was a result of Shackleton's boundless optimism and commanding personality. He led by example, getting involved in all entertainments and activities, even at one stage in late May consenting to shave his head when, in a moment of madness, the rest of the crew shaved theirs. Shackleton was also quick to stamp on any signs of conflict among the crew and when a delegation of men reported the bullying behaviour of John Vincent, the tough and imposing bo'sun, Shackleton swiftly disciplined Vincent by demoting him.

In the dark months of the Antarctic winter the temperature dropped to below −20° C and the pack-ice surrounding the ship began to shift ominously. Close to the ship, heavy pressure from the continuous force of the pack-ice forced huge blocks of ice into tall towers over 4.5 metres high. The sound of the ice as it shrieked and cracked unnerved the men, reminding Shackleton of the fragility of their position. This heavy pressure lasted for several days until calm returned by the middle of June.

ABANDON SHIP

At the beginning of July the hours of daylight started to increase and the men's thoughts turned to the prospect of freedom from the ice. Since becoming trapped they had drifted 1072 kilometres and were moving towards open water. However, on 13 July a gale suddenly turned into a harsh blizzard, and the worsening weather conditions increased the pressure in the pack-ice surrounding and squeezing the *Endurance*.

Shackleton was in his cabin discussing the perilous situation with Worsley and Wilson when the blizzard struck. Worsley later recalled, 'The wind howled in the rigging and I couldn't help thinking it was making just the sort of sound you would expect a human being to utter if they were in fear of being murdered ... "The ship can't live in this, Skipper." Shackleton said, "it is only a matter of time ... What the ice gets, the ice keeps.' Worsley was upset by Shackleton's cold logic and still believed that the *Endurance* could survive. However, Shackleton and Wild had by now accepted the eventual fate of the ship, although they kept the morale-sapping news secret from the rest of the crew.

Shackleton began to prepare for the inevitable evacuation. The decks were cleared and hourly watches established through the night. The mounting pressure of the ice shook the *Endurance* violently and the sides of the ship were forced inwards, buckling the seams and damaging the rudder. The ship was in danger of being destroyed totally, but miraculously the pressure suddenly stopped and calm returned to the pack-ice, although the changed icy landscape of ice-blocks and **hummocks** indicated the *Endurance*'s lucky escape.

KEYWORD

Hummock: a pile or ridge of ice.

The stay of execution, however, was not permanent. On 1 September another wave of pressure bore down on the *Endurance* and, for three long days, mercilessly squeezed the ship. McNeish, the ship's carpenter, wrote 'There were times when we thought it was not possible that the ship would stand it'. Throughout September, the instability of the ice produced powerful tremors that shook the *Endurance*. All through this anxious time Shackleton was a pillar of calm authority and sought to reassure the crew.

On 15 October though, the agony of the *Endurance* entered its final act. The ship broke free of the ice into a narrow lead of open water and spirits on board rose at the prospect of escape. However, on 17 October the pack-ice closed around the ship like powerful jaws, forcing the *Endurance* over onto her side. For days the crewmen battled to right the

Figure 5.1 The *Endurance* trapped by the pack-ice

ship but on 23 October the *Endurance* was crushed from three angles – across the **bow** and on both sides. The ship's **sternpost** was irreversibly damaged by the immense pressure and sea-water began to pour into the ship.

> **KEYWORDS**
>
> **Bow:** front of a ship.
>
> **Keel:** the bottom of a ship.
>
> **Sternpost:** the post at the rear of a ship.

Reacting to the crisis, Shackleton gave the orders to man the pumps but the water continued to flood the stricken ship. Other crew members desperately collected stores and food in preparation for the abandonment of the ship. Worsley raided the ship's library gathering maps and charts that would show a route to safety. For days Shackleton and his men battled in vain to stem the ship's leaks as the roar of the ice pressure and the screeching of the *Endurance*'s death throes filled the air.

On 27 October the pressure of the ice reached a crescendo and, when an ice-floe ripped out the ship's **keel** and water began to flood in, Shackleton gave the order to abandon ship. Shackleton wrote, '… no ship built by human hands could have withstood the strain. I ordered all hands out on the floe.' As the supplies that had been prepared were moved onto the ice floe, Shackleton stayed for a moment on board the *Endurance*, his emotions in turmoil at the loss of his brave ship, 'To a sailor his ship is more than a floating home,' he wrote. 'I cannot describe the impression of relentless destruction … The floes, with the force of millions of tons of moving ice behind them, were simply annihilating the ship.'

Shackleton was the last man to leave the *Endurance*. He assembled his exhausted and frightened men on an ice floe 110 metres from the ship, a hastily constructed camp littered with supplies. Their lives were now in Shackleton's hands and he was determined not to fail them.

✳ ✳ ✳ SUMMARY ✳ ✳ ✳

- After Amundsen and Scott reached the South Pole, Shackleton made plans to cross the Antarctic continent and obtained a specially built ship which he named the *Endurance*. He recruited Frank Wild as his second-in-command and Frank Worsley as ship's captain.

- The *Endurance* sailed from London on 1 August 1914 on the eve of the First World War. On his arrival at South Georgia, Shackleton was warned by the whalers that the pack-ice in the Weddell Sea was dangerously heavy. The *Endurance* sailed for Antarctica in December 1914 and soon encountered thick pack-ice.

- In January 1915 the *Endurance* neared Antarctica but the pack-ice closed around the ship trapping the expedition. The *Endurance* drifted with the Weddell Sea currents and Shackleton set up a strict routine to combat boredom. From July through to October, the pressure of the pack-ice squeezed and damaged the ship.

- On 27 October an ice-floe ripped out the *Endurance*'s keel and water flooded the ship. Shackleton ordered his men to abandon ship and set up camp on the ice-floe.

Life on the Ice

6

After the crew abandoned the *Endurance* they spent a restless night on the ice-floe huddled in tents, as the ice continued to crush the ship. Shackleton stayed on watch, the safety of the men his major concern.

On the morning of 28 October 1915, their first day on the ice, Shackleton gathered the men together. All hope of crossing the Antarctic continent had vanished and Shackleton simply outlined their new challenge – to get home. This challenge was a massive one; in the ten months since the *Endurance* had first become trapped in the ice, it had drifted 2080 kilometres. The nearest inhabited land, the island of South Georgia, was over 1600 kilometres away across the menacing pack-ice and hundreds of kilometres of stormy ocean.

Shackleton's plan was to march westwards over the pack-ice to Snow Hill Island before the ice began to break up. Three lifeboats had been salvaged from the *Endurance* and two of these, the *James Caird* and the *Dudley Docker*, were brought on the march. This was a precaution in case the ice began to crack up and they needed to take to the water. Sledges were fitted to the bottom of the lifeboats to allow the crew to haul them along the ice.

KEY FACT

Snow Hill Island was 500 kilometres away to the west. Shackleton knew there was a hut stocked with emergency supplies there, which he had ordered for a mission to rescue the Swedish explorer Otto Nordenskjold in 1902.

The march began on 30 October. Shackleton, accompanied by expedition members, Wordie, Hussey and Hudson, led the way. Their job was to break down any ice blocks that obstructed the route. Behind them came the rest of the crew, man-hauling the lifeboats which, when loaded with rations and essential equipment, weighed as much as a ton each.

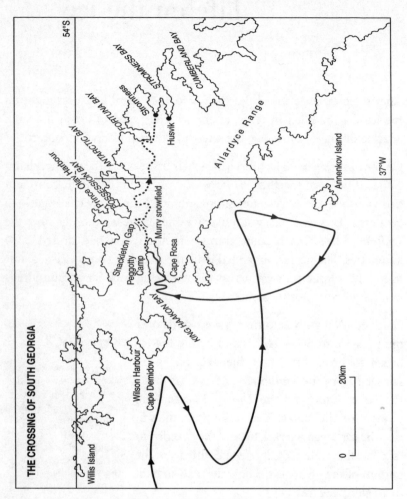

Figure 6.1 The drift of the *Endurance* in the Weddell Sea

The surface of the ice did not help the crew's progress, as the expanse was covered with huge **pressure ridges**. By the end of the third day, they had only covered one-and-a-quarter miles. Shackleton decided to abandon the march.

KEYWORD

Pressure ridge: a place where colliding ice-floes have thrown up huge slabs of sea ice.

OCEAN CAMP

The men set up camp on a stable part of the floe to wait for the ice to break up. Only then would they be able to launch the lifeboats and sail to land. The camp was named Ocean Camp, a light-hearted reminder of their dangerous situation. The third lifeboat, the *Stancomb Wills*, and supplies they had left behind were brought back to Ocean Camp from the wreck of the *Endurance*.

A routine was established to keep up the men's morale. At 8.30 a.m. breakfast was served, a meal consisting of fried seal, **bannocks** and tea. Shackleton organized the crew into teams of men, who spent the mornings hunting for seals

> **KEYWORD**
> Bannocks: humps of baked dough.

and penguins to add to the stock of food. Lunch was served at 1.00 p.m., and after this the men spent the afternoons as they wished; walking or reading one of the few books rescued from the *Endurance*. Supper was at 5.30 p.m. – hoosh with a mug of cocoa. The crew then retired to their sleeping bags for the night. Shackleton ensured that watches were set up throughout the night to warn the camp if the ice suddenly began to break up.

> **KEY FACT**
> Seals were not only a source of food, but also provided the crew with fuel for cooking. The thick layers of fat called blubber that insulate a seal's body contain oils and 75 square centimetres (one square foot) of seal blubber will burn for several hours.

THE *ENDURANCE* SINKS

Mid-November saw the beginning of the Antarctic summer and an increase in temperature. The snow on the ground in Ocean Camp turned to slush, soaking the men's sleeping bags. In the crowded tents, tempers frayed as the temperature rose. To battle against the growing unease in the camp, Shackleton made sure that the men were kept busy and ordered the ship's carpenter, Harry McNeish, to ready the lifeboats for the open sea.

As a reminder of their desperate situation, the wreck of the *Endurance* remained in sight a few kilometres in the distance. However, on

21 November Shackleton's earlier words, 'what the ice gets the ice keeps' were realized as the *Endurance* finally sank. From the camp's lookout, Shackleton called out to the men, 'She's going boys' and they all gathered for a last look at the doomed ship. At first the *Endurance*'s stern rose high in the air and then, with one swift dive, the ship vanished beneath the ice.

The men on the ice had mixed feelings. To some, the sight of the wreck of the *Endurance* had been a source of depression, but for Shackleton the loss of the ship was painful. In his diary he simply noted, 'At 5.00 p.m. she went down ... I cannot write about it.' The last symbolic connection with the outside world had gone.

During their time at Ocean Camp Shackleton charted the direction of the ice-floe's drift. In December the floe began to drift eastwards away from land. To counteract this, Shackleton decided that they should again attempt the march west. This decision was motivated by his concern about the mood of the men. Some of the sailors believed that since the *Endurance* had sunk they would no longer be paid, and they created an atmosphere of dissatisfaction. Shackleton discussed the situation with Frank Wild, his second-in-command, and they agreed that the march could help to restore morale.

A MUTINY AVERTED

However, in the camp, feelings about the march were not so positive. As it was the height of the Antarctic summer, the men feared that the condition of the ice underfoot would be even more treacherous than before. Shackleton decided that the march would take place at night in the hope that the ice would be firmer.

Christmas Day was celebrated on 22 December before the trek started and Shackleton allowed the men to eat as much food as they wished. This had the effect of lifting the men's spirits, as well as physically preparing them for the journey ahead. The next day the march began. Once again the lifeboats were mounted on sledges and the men, harnessed like dogs, dragged them over the ice. At times the going was terribly soft, and the men became soaked by the slushy ice and exhausted by the effort of marching at night. After four days they had managed only eight kilometres.

On 27 December, one of the men hauling the boats, Harry McNeish, stopped and refused to go on. Frank Worsley, the officer in charge of the lifeboat haulers, ordered him to continue but McNeish refused to obey. Shackleton was summoned to deal with the situation. McNeish argued with Shackleton that since the *Endurance* had been abandoned the **Ship's Articles** no longer applied. This meant that he did not have to obey an officer's orders and could refuse even Shackleton's commands. Shackleton knew that McNeish was not only voicing his own opinions but the concerns of others in the crew. If this rebellion was not crushed immediately the men's trust and belief in him would be shattered.

> **KEYWORD**
>
> Ship's Articles: the name given to the contract that sailors agree to uphold when they take service on a ship.

Shackleton's response was swift and decisive. He gathered together the crew and calmly read aloud the Ship's Articles, '… the crew agree to conduct themselves in an orderly, faithful, honest and sober manner … disobedience to lawful commands will be legally punishable.' He informed the crew that, contrary to usual practice, Ship's Articles had not ended with the loss of the *Endurance* and they would be paid until they reached a safe port. This calmed the men's worries and isolated McNeish.

To ensure McNeish's future obedience Shackleton spoke to him alone and hinted that if he continued to defy orders he would be shot. Behind Shackleton's cold threat lay a real fear about the impact of McNeish's defiance on the morale and mental well-being of the rest of the crew.

Shackleton's ability to face new challenges was about to be tested further. By 28 December, conditions on the march had further deteriorated; at times the ice was so thin that the sledges carrying the lifeboats broke through to the seawater beneath. Shackleton put the safety of his men above any loss of pride and abandoned the march. He recorded in his diary, '... decided to retreat to more secure ice: it is the only safe thing to do.' At the end of December a new base was established on the ice – Patience Camp.

LAUNCH OF THE LIFEBOATS

For three long months, Shackleton and his men waited. As their food supplies dwindled, the men spent more and more time in their sleeping bags. Shackleton sadly ordered the shooting of the remaining dogs. The food that they consumed had become too precious to the men, as even the **pemmican** from the dogs diet was eaten by the crew.

On 7 April, land was finally sighted – Clarence Island and Elephant Island – 95 kilometres to the north-west. The floes began to crash against each other, grinding away the ice on which they lived. Shackleton wrote, 'our home was being shattered under our feet.'

> **KEY FACT**
>
> Clarence Island and Elephant Island are the northernmost islands in the South Shetlands chain. They were Shackleton's last chance of making land, as beyond them lay the Atlantic Ocean.

As the ice began to crack and break up, Shackleton readied the boats. He took charge of the *James Caird*, the largest of the lifeboats, along with Wild and 11 other men. Worsley was placed in charge of the *Dudley Docker*, and Hudson and Crean given command of the *Stancomb Wills*. At midday on 9 April, the ice opened up allowing them to launch the lifeboats.

Since the loss of the *Endurance*, Shackleton had led the men safely through the trials of life on the ice. Now, he and the rest of the crew relied upon Frank Worsley, the captain of the *Endurance*, to guide them to land and safety. Worsley had to navigate in stormy seas and high winds. Poor weather conditions meant that he had to rely on **dead reckoning** to guide him.

> **KEYWORD**
>
> Dead reckoning: a method of navigation that works out where a boat's position will be at a certain time, so long as it maintains a specific speed and direction.

ELEPHANT ISLAND

On the first day in the water the boats had to make their way through the dangerous pack-ice. On several occasions the floes closed in on the boats threatening to crush them. It was with relief that the men camped on a flat ice-floe, measuring 60 × 30 metres, at the end of the first night. The men set up tents and, with the aid of the stove, had a hot meal.

As the men slept, Shackleton reported a 'feeling of uneasiness made me leave my tent about 11 p.m. that night.' As he surveyed the quiet makeshift camp, the floe was lifted by a swell and the ice cracked under his feet. The crack ran through the sailor's tents dumping Ernest Holness, still in his sleeping bag, into the water. Shackleton reached into the crack and pulled Holness out, moments before the edges of the ice slammed back together. An anxious night followed before the boats could be launched in the morning.

On the second day in the boats, the men continued to row through the pack-ice in the belief that they were heading westwards towards land. However, three days into the journey, a sighting of the sun allowed Worsley to calculate their position using his navigational tables. These revealed that a strong easterly current, combined with the effort of navigating a route through the pack-ice, had misled the men's sense of direction and forced them 50 kilometres to the east. Shackleton played down this news, which could have had a devastating effect on the morale of the crew.

A terrible night in the boats wore down the resolve of the men. In the storm-driven seas, temperatures dropped to below 7° C. As snow showers fell, the boats became caked in thick ice that had to be hacked off with axes before its weight sank them. Shackleton could see the physical and mental state of the crew deteriorating before his eyes. He could no longer afford to be cautious; he had to race to save the lives of his men. By midday on the fourth day, the boats finally emerged from the pack-ice and, with the wind behind them, surged towards Elephant Island.

By Thursday 13 April, the fifth day in the boats, Shackleton feared that the crew had begun to suffer from the effects of exposure. Any delay in reaching land and shelter could cost the men their lives, and so Shackleton refused permission to pause in their journey, even to cook food. This meant that the men had to chew pieces of raw seal meat for nourishment, and drink the blood to quench their thirst. However, the saltiness of the meat intensified their thirst and increased the men's suffering aboard the tiny boats.

> **KEY FACT**
>
> Exposure to the elements is a great threat to survival. It can result in sunburn, frostbite and most seriously hypothermia – where body temperature becomes dangerously low, sometimes causing death.

Elephant Island came into view on the Friday; Worsley had miraculously calculated their direction perfectly. However, storm-driven waves pushed the tiny boats away from the coast of Elephant Island. As darkness fell, the chance of landing that day had gone. Shackleton was forced to spend another night at the tiller of the James Caird. He had not slept since they had left Ocean Camp, six days earlier.

Eventually, at dawn on 15 April, the storm eased and the boats were able to sail around the coast of Elephant Island. The coast was bleak with sheer cliffs preventing a landing. Finally, a sheltered bay was found and the boats were pulled up on the shore. The men stumbled up the beach – the first land they had stood upon since 5 December 1914, 497 days earlier. The journey had taken a terrible toll; many of the men were suffering from the effects of exposure, whilst others wandered around the beach dazed and crying, overcome by the mental strain of the journey. Shackleton, with the help of Worsley's navigational skills, had led the crew to dry land. However, they were still far from safe. Elephant Island was uninhabited and nobody in the outside world knew that they were there. South Georgia and civilization was over 1,125 kilometres away across the stormy Atlantic Ocean.

✸ ✸ ✸ SUMMARY ✸ ✸ ✸

- On 27 October 1915, Shackleton ordered the crew to abandon the *Endurance*. After a failed attempt to march over the ice to Snow Hill Island, the men set up Ocean Camp in early November. Seals and penguins were hunted for food.

- In December the ice-floe began to drift eastwards away from land. Shackleton and the crew attempted to march westwards again. During the march, Harry McNeish, the ship's carpenter, refused to obey orders and forced Shackleton to assert his authority. Worsening weather conditions meant that the march was abandoned and Patience Camp set up.

- In the early months of 1916 the ice-floe drifted northwards towards the South Atlantic Ocean. On 30 March, Shackleton ordered the shooting of the dogs, as food supplies became dangerously low.

- On 9 April, the ice began to break up and the three lifeboats, the *James Caird*, the *Dudley Docker* and the *Stancomb Wills*, were launched. After seven exhausting days at sea the lifeboats finally landed safely at Elephant Island, an uninhabited island 1,125 kilometres from civilization.

Into the Unknown

On the desolate beach on Elephant Island, Shackleton's first emotion was relief, 'Thank God I haven't killed one of my men,' he confided to Wilson soon after landing. However, this relief was soon tempered by several disturbing signs on the beach. High tide marks and fallen boulders from the cliffs above indicated that the beach was no place of safety.

After a night's rest, Wild led a party out in the *Dudley Docker* in search of a safer camp. Eleven kilometres along the craggy north coast of the island Wild spotted a **spit** of land that seemed to offer a safe haven. Once Wild returned with the news, Shackleton ordered the men to load the boats and head for this newly discovered camp. As they set out again on 17 April, storms battered the small boats as they edged their way along the coastline. At one stage Worsley, in the *Dudley Docker*, was almost blown back out to the open seas, but eventually all three boats arrived safely at their new camp by nightfall.

> **KEYWORD**
> Spit: a point of land stretching into the sea.

LIFE ON ELEPHANT ISLAND

In honour of its discoverer, the camp was named Cape Wild and it more than lived up to its name. Shackleton described it as '… rough, bleak, and inhospitable – just an acre or two of rock and shingle, with the sea foaming around it.' However, to the west there was a glacier which supplied drinking water and the spit was also home to many penguins and seals which would provide invaluable food and fuel. After supper and a night's sleep, Shackleton ordered the crew to kill all the penguins that they could catch in order to stockpile food before the birds migrated north for the winter.

Shackleton revealed his plans to Wild and Worsley, his two most trusted companions and outlined their roles, '… I had … to tell Wild that he must stay behind, for I relied upon him to hold the party together while I was away ... I determined to take Worsley with me as I had a very high opinion of his accuracy and quickness as a navigator.' Two days after the arrival at Cape Wild, Shackleton outlined his plans to obtain rescue to the rest of the crew. There was no chance of any search party looking on the isolated island for them, so the only option was another boat journey in search of relief. He announced that he was heading off immediately in the *James Caird* for South Georgia to fetch help.

> **KEY FACTS**
>
> South Georgia was over 1,125 kilometres away, over the Southern Ocean. Although the Falkland Islands were closer, situated 865 kilometres to the north, South Georgia was to the north-east in the direction of the prevailing winds.
>
> The *James Caird* was 6.6 metres long and, at its widest, measured two metres across.

Shackleton asked for volunteers to accompany him on this perilous mission. To many of the men who had barely survived the escape from the ice alive, the idea seemed like certain suicide. McNeish, the carpenter, was one of the few volunteers and Shackleton decided to take him as his skills would be essential aboard the small wooden boat. He was also mindful of McNeish's earlier mutiny and worried on the effect that further rebellion could have on the fragile community at Cape Wild. This was also reflected in his selection of Vincent, the bo'sun previously demoted for bullying behaviour. In addition to these men Shackleton also selected the optimistic McCarthy and the hardy Crean.

To maximize their chances of survival, Shackleton ordered McNeish to make several improvements to the *James Caird*. To prevent the small boat from being swamped by the fierce seas, the boat was decked in, with the resourceful McNeish using wood taken from sledges and packing cases.

THE STORMIEST SEAS ON EARTH

Confidence among the men at Cape Wild was low, as 95 k.p.h. winds swept down over the cliffs and through their makeshift camp, and it was just past midday on Monday 24 April that Shackleton, Worsley and the rest of the tiny crew set sail in the *James Caird*. They raced away from the shore, hurrying to avoid the pack-ice that was gathering off the coast.

Hurley photographed the poignant moment as the remaining crew cheered their brave leader on his way. Ahead of Shackleton lay over 1,125 kilometres of the stormiest and most unpredictable seas on the planet.

With Worsley at the helm, the *James Caird* headed north into the open seas; Shackleton content to put his faith in Worsley's excellent seamanship. At the end of the first day out from Elephant Island Shackleton ordered Crean, McNeish, Vincent and McCarthy down below to rest, whilst he and Worsley guided the boat through the night. The conditions were gruelling with the *James Caird*'s makeshift cockpit offering scant protection against the waves, forcing Shackleton and Worsley to huddle together for warmth, talking through the night as they became soaked to the bone. At the end of their first 24 hours they had travelled 72 kilometres from Elephant Island, but the weather began to worsen. A force 9 gale created waves, six metres high, that mercilessly tossed and shook the tiny *James Caird*. To prevent the boat being swamped, Shackleton set up a series of watches with the men split into two teams, one led by himself, the other by Worsley. These teams worked in four-hour shifts to steer the boat, man the sails and desperately bale out the near overwhelming sea waters that flooded in with every wave.

However, if Worsley was not able to navigate their passage to South Georgia, the seaworthiness of the *James Caird* would be for nought and it was only on 26 April, three days out from Elephant Island, that Worsley was able to take an observation using a **sextant**. In the pitching seas this simple act meant dicing with death. Using the **chronometer** Worsley calculated that their position was 59° 46′ S, 52° 18′ W, 205 kilometres distant from Elephant Island. Using these rough estimates and making allowances for the pitching of the boat and the accuracy of the chronometer, Worsley had to plot their course to South Georgia, 965 kilometres to the north-east. If they missed the small island they would be swept out into the South Atlantic with no hope of survival.

KEYWORDS

Sextant: an optical instrument used for measuring the distance between the sun and the horizon as a means of navigation.

Chronometer: a special type of clock used in navigation.

The potential for disaster weighed heavily on Shackleton's mind and he did everything he could to protect those on board. His Antarctic experiences had taught Shackleton that the three essentials for survival were food, sleep and warmth. With warmth an impossibility and sleep a snatched and fitful luxury, Shackleton made sure that the crew did not want for food. Crean acted as cook and used a small primus stove to prepare hot meals of sledging rations which the men consumed twice a day. Shackleton supplemented this with biscuits, sugar lumps and mugs of hot dried milk.

The freedom from hunger was little consolation for the perpetual hammering that the *James Caird* took as the journey wore on. Six days into the voyage, Worsley was able to take a second sighting of the sun and calculated that they had covered a third of the distance to South Georgia. On the seventh day, a harsh gale created disorientating conditions and the boat began to ship water to a dangerous extent. Shackleton was forced to put out a **sea anchor** to help steady the boat.

> **KEYWORD**
>
> Sea anchor: a cone-shaped canvas bag which is attached by a line to the bows of a ship and acts as a drag in the water, forcing the ship to keep its head to the wind.

BEYOND THE LIMITS OF ENDURANCE

The cold on board the *James Caird* was so intense that on 1 May, after a disturbed sleep, his first for several days, Shackleton discovered that the boat had become iced over. The ocean spray had frozen to the deck and sides of the vessel, in places almost a third of a metre thick. To prevent the overburdened boat from capsizing the men desperately hacked at the ice with an axe. Worsley described the painful process with his usual fortitude, 'The boat leaping and kicking like a mad mule and a ... thick slippery casing of ice over her ... First you chopped a handhold; then a kneehold and then went on chopping ice for dear life while an occasional sea leapt over you.' Their battle against the ice was constant, as any lapse in concentration could send the *James Caird* to the bottom of the ocean.

For Vincent, the intense physical demands became too much and he suffered a mental collapse. The sheer effort of their exertions was beginning to tell on all of the men, with signs of exposure affecting McNeish in particular. Shackleton continued to work hard to protect his men as Worsley noted, 'He [Shackleton] ... has his finger on our pulse and at the psychological moment orders a hot feed. This ... saves ... our lives.'

On the tenth day out the weather improved sufficiently to allow Worsley to take a sighting of the sun and calculate their position. He worked out that they had covered 710 kilometres and were less than 480 kilometres from South Georgia. This pleasing news, in combination with the kinder weather, temporarily raised the men's spirits.

It was only a temporary respite. On 5 May, a gale accompanied by snow showers soaked and threatened to flood the tiny boat. The worst, however, was still to come as, at midnight, the mighty Southern Ocean revealed its true strength. Shackleton described the sight that greeted him from his position at the tiller of the boat:

> '... the white crest of an enormous wave. During twenty-six years experience of the ocean in all its moods I had not encountered a wave so gigantic ... shouted, "For God's sake, hold on! It's got us!" Then came a moment of suspense that seemed drawn out into hours ... We felt our boat lifted and flung like a cork in a breaking surf. We were in a seething chaos of tortured water, but somehow the boat lived through it, half-full of water, sagging to the dead weight and shuddering over the bow. We bailed with the energy of men fighting for life...'

Miraculously they survived and after ten minutes of frantic bailing rode out the storm. After surviving the gale, they made the painful discovery that their drinking water had become spoiled with seawater. The spectre of thirst was another foe that Shackleton had to defeat.

SOUTH GEORGIA, BUT NOT SAFETY

On 7 May, Worsley was able to take another observation and, using dead reckoning, calculated that they were within 145 kilometres of South Georgia. Sightings of small birds and pieces of kelp hinted at the closeness of land and at 12.30 p.m. on 8 May the barren and uninhabited south coast of South Georgia was sighted. So close to salvation, the men now faced the greatest danger. With the light failing, Shackleton had a difficult decision to make. Worsley wanted to attempt an immediate landing, but Shackleton's knowledge of sea conditions warned him of the dangers of landing in fraught circumstances. Feeling that they had come too far only to risk their safety for a lack of caution, Shackleton ordered Worsley to stand the boat off for the night. After a painful night's battering by the treacherous coastal seas, Worsley guided the *James Caird* along the mountainous coast. At times the winds reached hurricane force and threatened to wreck the small boat on the half-submerged rocks; crashing waves, shrieking winds and thick mist that shrouded every-thing in a murky gloom made landing an impossibility and Worsley was again forced to pull the boat away from the shoreline.

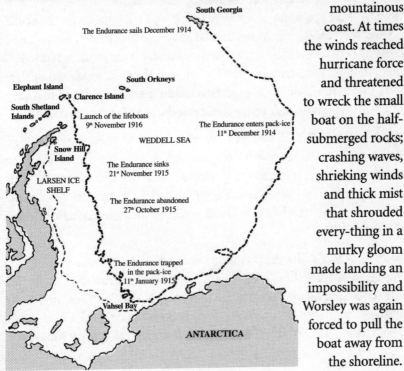

Figure 7.1 The route across the uncharted interior of South Georgia

Another exhausting day at sea had passed and the men were at the point of collapse; they had not drunk any water for over two days and Shackleton decided that they had to attempt a landing at King Haakon Bay on the uninhabited south side of the island. Under Worsley's skilful command and Shackleton's watchful eye, the *James Caird* was able to fight her way into the narrow **fjord** and finally land on a small beach at the tip of the cove.

> **KEYWORD**
>
> Fjord: a narrow inlet of sea.

It was 10 May 1916 and they had been aboard the *James Caird* for 17 days. The first man ashore was Shackleton; he had defied the mighty Southern Ocean and led his men to the promised land of South Georgia.

Their first night was spent sheltering in a small cave at the base of a cliff. In the morning Shackleton considered the situation. The *James Caird* wouldn't take the strain of a voyage to the whaling stations on the north side of the island, which meant that they had to attempt an arduous crossing on foot over the uncharted mountain ranges of the interior. His men, though, were in no fit state to attempt the trek, weakened as they were by the strain of the voyage. Shackleton therefore ensured that they built their strength back up by resting in the cave for several days.

While resting, Shackleton and Worsley explored their surroundings. From their vantage point they spotted a possible route into the mountains, a pass through the glaciers at the other side of the bay. Four days after landing, the men once again loaded the *James Caird* and made their way across the bay to a small beach located underneath the pass they had sighted.

McNeish, McCarthy and Vincent were still too weakened from the voyage to attempt the deadly crossing. So, after constructing a shelter for the remaining men by placing the upturned *James Caird* against a rocky outcrop and fashioning walls and a door using rocks and grass, Shackleton, Worsley and Crean set off on the morning of 19 May.

THE CROSSING OF SOUTH GEORGIA

The challenge that lay ahead of them was immense. Only 32 kilometres separated them from the manned whaling station at Husvik. However, the Antarctic winter had set in, worsening the already unpredictable South Georgian weather with its mixture of gales, snowstorms and dense mists. Shackleton's only hope of success was constant travel, day and night, with just the moon to guide their path across the unmapped mountains and glaciers of the interior. It was a final step into the unknown.

Modern mountaineers would find the conditions on South Georgia intimidating, but for Shackleton, Worsley and Crean, with their rudimentary equipment and worn through clothing, it was the ultimate challenge. At the summit of the pass Shackleton recorded the sight that lay in front of them: 'High peaks, impassable cliffs, steep snow-slopes, and sharply descending glaciers could be seen in all directions.' Roped together for safety, Shackleton led the way while Worsley called out directions as he plotted their course using their only compass. A routine was swiftly established, with the party stopping every quarter of an hour to prevent the onset of exhaustion. Regular breaks were also made for snacks of biscuits and sledging rations.

All of the men were operating at the outer limits of endurance and the disorientating terrain, in combination with their complete exhaustion, led the party down several blind alleys. Time and again they descended from the peaks they had scaled in the expectation of finding a safe path only to be met by sheer precipices.

By 8.00 a.m. on the morning of 20 May, they reached the northern side of the island. They emerged, however, far to the west of their intended destination of Husvik and, as there was no safe route along the coast, they were forced to turn back to look again for the right path through. The men tracked wearily south-eastwards, only to be confronted by an undulating mountain ridge topped by five distinct peaks. Shackleton could see four passes between the peaks and, with no time to scout out the safest route, he was forced to make an instant decision. Trusting his instinct he chose the southernmost pass which appeared to be the lowest and therefore the least demanding on his weary men.

As they ascended the pass the men sank up to their knees in the snow, which made their progress slow and tiring. It took four hours to reach the head of the pass and the view that greeted them was a morale-sapping one. As Shackleton later recalled, 'I looked down a sheer precipice to a chaos of crumpled ice 1,500 feet below. There was no way down for us.' Again, the men retraced their steps before climbing again to attempt the second pass. The day that had started so brightly was now fading and thick fog began to bar their path. At the peak of the second pass further heartbreak awaited them. From their vantage point, no safe path down could be sighted. Worsley expressed a desire to push on and risk a descent but Shackleton, with the overall responsibility for all the men left behind, counselled caution. The disappointment would have broken a lesser man, but Shackleton merely led the party back down in the fading light and up the third pass, mindful of the danger of ending the day without progress. At the head of the third pass Shackleton spied an indistinct path downward through the fog and encroaching darkness. 'We'll try it', was his decision and using an **adze** to cut steps into the ice he led the team downwards.

> **KEYWORD**
> Adze: a climbing axe.

The strain of cutting a path down was both tiring and time-consuming. With the night closing in and the danger of the weather worsening in such an exposed position, Shackleton was forced to take an enormous risk. He decided that they would toboggan down the ice slope, even though the bottom was out of sight in the gloom below. The men each coiled up a portion of the rope that they had been using to link themselves together and sat on it, so it acted as a sled. Shackleton sat at the front with Worsley close behind, while Crean brought up the rear. Worsley later related the excitement and tension as they launched themselves down the slope like a bedraggled bobsleigh team: 'For a moment my hair fairly stood on end. Then quite suddenly I was grinning! I was actually enjoying it.' The men sped down the slope until the gradient levelled out, and they ended their ride in a snowbank at the base of the slope. They had descended 450 metres in three minutes.

STROMNESS BAY

As the moon rose in the chill sky, Shackleton encouraged the men onwards. Without the shelter of a tent in the bleak South Georgian interior, the risks of resting for any length of time were too great. They had been marching solidly for 24 hours and, even though Shackleton maintained a routine of meals and brief rest stops, the toll of their exertions began to tell on the men. Again they descended towards the coast in the mistaken belief that they were nearing Stromness Bay – a whaling station – until crevasses warned them that they were actually climbing down a glacier towards the ocean.

As they turned back once more to regain the height that they had lost, Shackleton allowed Crean and Worsley to rest while sheltering behind a rocky outcrop. Recognizing the risks of exhaustion he kept a watchful eye on the men and after five minutes shook them awake and told them they had slept for half an hour. As they continued on their way, the breaking dawn revealed to Shackleton a peculiar rock formation that he recognized from his stay at Stromness Bay 17 months earlier. To confirm that the vision was not some exhaustion-induced mirage, at 7.00 a.m. the whistle summoning the whalers to work shrilled out across the last remaining kilometres.

After a swift breakfast Shackleton led the men on the final stretch of their epic journey. They had been travelling for nearly 30 hours without a proper rest and, as another steep slope lay in their path, the last kilometres would prove to be some of the hardest. Drawing on their final reserves of energy they managed to climb the Stromness mountain range and from the peak looked down on the whaling station beneath them.

The descent down to Stromness Bay presented a final hardship on the weary survivors. As they blundered down a steep and narrow valley, which forced the men to wade down a freezing stream, they were met by a nine metre waterfall. Using the rope, they abseiled down and finally found themselves on level ground. After tumbling over frozen marshlands they reached the whaling station at 4.00 p.m.

On the station's wharf, Shackleton approached the foreman, Matthias Anderson and asked for the manager, Captain Anton Anderson. The foreman informed Shackleton that Thomas Sorlle, whom Shackleton had befriended when the *Endurance* had docked at South Georgia, was the new station manager. Anderson led the strange group to Sorlle's house, trailed by a gang of curious whalers. Sorlle was told by his foreman 'There are three funny-looking men outside, they say they know you.' When Sorlle emerged from his house Shackleton calmly asked him, 'Do you know me?' Sorlle replied hesitantly, 'I know your voice.' Shackleton informed him 'My name is Shackleton.'

The man whom the entire world presumed dead had by a miracle of skilful leadership and supreme bravery returned from the most remarkable adventure of all time. He had achieved the impossible. Now all that remained was the rescue of his men.

* * *SUMMARY* * *

- The men set up camp at Cape Wild on Elephant Island and, on 24 April 1916, Shackleton set out in the *James Caird* accompanied by Worsley, Crean, McNeish, McCarthy and Vincent for South Georgia, 1,125 kilometres away over the Southern Ocean.

- Conditions on the 17-day voyage were arduous with the men fighting against gales, giant waves and the effects of exposure. When they reached South Georgia stormy weather conditions forced Shackleton to land the boat on the uninhabited southern side of the island.

- Shackleton, Worsley and Crean crossed the uncharted interior of South Georgia, marching non-stop and traversing glaciers and mountains, to reach the manned whaling station at Stromness Bay.

- Shackleton and his team reached the manned whaling station at 4.00 p.m. on 21 May 1916. He knew the station manager, Thomas Sorlle, from a previous visit to South Georgia, but Sorlle didn't recognize Shackleton with his matted beard and tattered clothing.

Shackleton's Legacy 8

His epic journey was at an end, but Shackleton still could not rest. The entire adventure would be worthless if he did not secure the rescue of his stranded crew. Shackleton related the tale of their escape from the ice to Sorlle, outlining the fate of the crew marooned on Elephant Island and at King Haakon Bay and Sorlle swiftly organized a whaling vessel, the *Samson*, to rescue Vincent, McCarthy and McNeish. After a welcome night's rest, Worsley accompanied the crew of the *Samson* to find the remaining members of the boat party.

Shackleton remained at Stromness to begin planning the rescue of the men stranded on Elephant Island. A large steam whaler, the *Southern Sky*, was anchored at Husvik for the winter and Sorlle arranged for Shackleton to be loaned the ship. Many of the whaling station men volunteered to crew the ship, including a hardy Norwegian sailor, Ingvar Thom, whom Shackleton made captain.

Before he departed on the final leg of his rescue mission, Shackleton and his fellow adventurers were accorded a celebratory reception by the whalers of Stromness. To a captivated audience Shackleton recounted their journey from Elephant Island. The tale was so amazing that one experienced sailor was moved to announce that '… never had he heard of such a wonderful feat of daring seamanship … he felt it an honour to meet and shake hands with Sir Ernest and his comrades … "These are Men!"'

RESCUE FROM ELEPHANT ISLAND

On 23 May, Shackleton sailed in the *Southern Sky* for Elephant Island. It had been a month since he had left the island and the welfare of the men he had left there haunted his every thought. Crean and Worsley accompanied Shackleton on this rescue bid, while Vincent, McCarthy and McNeish returned to England. The Antarctic ice continued to dog

Shackleton's progress as three days into the voyage the *Southern Sky* encountered impenetrable pack-ice ice. Although the ship tried to break through, she was eventually forced to head for the Falkland Islands, a British territory off the coast of South America, to seek assistance.

With no wireless or cable on South Georgia, news of Shackleton's survival had not reached the outside world, but on his arrival in the Falklands news of his exploits resounded around the globe. Even in the darkest days of the First World War, the miraculous reappearance of Shackleton was front-page news. However, when Shackleton contacted the Admiralty to request assistance with the rescue of the stranded men on South Georgia, he received a colder reception. In the middle of wartime, days after the **Battle of Jutland**, the navy could spare scant resources to help Shackleton. Back in Britain, a time-consuming refit of the *Discovery* was taking place to carry out Shackleton's relief, but it would be several months before she was able to sail. However, the Foreign Office asked the governments of Argentina, Uruguay and Chile whether they could spare any vessels to complete the rescue.

> **KEY FACT**
>
> The Battle of Jutland: a major Anglo–German naval battle which took place in European waters.

After many months of single-minded determination, Shackleton was now dependent on the actions of others. All through June, July and into August, frustration mounted upon frustration as no suitable ship was found. Shackleton, accompanied at all times by the steadfastly loyal Worsley and Crean, headed for Punto Arenas, a major Chilean port. After a plea to the naval commander there, Shackleton was loaned an ageing tug named *Yelcho*. Even though the boat could not survive in the pack-ice, Shackleton's desperation was such that he set out on 25 August for Elephant Island, blindly hoping that the seas around the island would be free of ice. With the weather staying fair the voyage was swift and the *Yelcho* encountered no pack-ice. By 30 August the ship closed in on Elephant Island and the familiar barren coastline loomed into view.

For the stranded men, the sight of the ship was a miracle; it had been 126 days since Shackleton had departed on his rescue mission aboard the *James Caird*. On Elephant Island, Wild had kept the crew together, both mentally and physically, but by this time food was running out. When a rowing boat from the *Yelcho* approached the shore the castaways cheered and waved in near disbelief as the figure of Shackleton was identified.

As soon as the boat reached the shore, Shackleton urgently enquired whether all the men were safe and well, and the news that all had survived unharmed brought a smile of relief to his lips. The stranded men quickly gathered together their few possessions, hurried on by Shackleton who was anxious to escape before the pack-ice had a chance to appear and they soon departed for Punta Arenas. Crowds of people lined the harbour there to greet the hardy adventurers and, eager to create a memorable entrance, Shackleton had instructed the men not to change their clothing or cut their beards and hair. After a tremendous welcome, Shackleton penned a celebratory letter to his wife Emily informing her of the successful rescue, 'I have done it … Not a life lost and we have been through hell.'

Yet for Shackleton, his mission was still not complete. News had reached him of the fate of the *Endurance*'s relief expedition who had laid the Ross Sea food depots for the aborted transantarctic crossing. In recognition of their loyalty and bravery, Shackleton ensured that he accompanied the Australian ship that carried out the rescue of the Ross Sea crew.

After the successful relief of the Ross Sea party Shackleton didn't immediately return to England, but set off on a lecture tour of America, recounting the tale of the *Endurance* expedition. With the United States a new participant in the

KEY FACT

The Ross Sea party endured a nightmare expedition. Their ship, the *Aurora*, had slipped her mooring and drifted out to sea, with pack-ice preventing her return to base and marooning the relief party on the Antarctic continent. Mindful of their role, the stranded crew members still managed to lay the food and fuel depots along Shackleton's planned transantarctic route. The men in the Ross Sea party, like the crew of the *Endurance*, suffered appalling strains and hardships, with three lives lost.

First World War, Shackleton's visit was an ideal opportunity to promote the war effort. American audiences were enthusiastic and hailed Shackleton as a giant among men for his courageous exploits.

On his return to Britain in May 1917, the public reaction to Shackleton was one of indifference, with his expedition viewed as a failure and his lectures, accompanied by Hurley's stunning photography, played to half-empty halls. In a land haunted by the casualties of war, Shackleton was not seen as a returning hero but an irrelevance.

Although at the age of 42 Shackleton was exempt from active service, the desire to prove his courage still burned brightly and he tried desperately to obtain a commission. The Government first sent Shackleton on a propaganda mission to South America, and then posted him to Murmansk, a north-western Russian port, as 'Staff Officer in charge of Arctic Transport'. The only comfort for Shackleton in this distant posting was the company of several old *Endurance* colleagues, including his ever-loyal deputy Wild.

THE *QUEST* EXPEDITION

When the war ended, Shackleton was again rudderless and life at home with his family was little consolation to a heart that craved the open spaces of Antarctica and the responsibilities of leadership. In a letter to his wife written in 1919, Shackleton revealed, 'Sometimes I think I am no good at anything but being away in the wilds.' This yearning for the polar regions manifested itself in his plans to explore the Arctic seas to the north of Canada. With financial help from a wealthy old friend, John Quiller Rowett, Shackleton obtained a ship, the *Goshawk*, which was renamed the *Quest*. As his plans for the expedition were made public, several of his *Endurance* crewmates pledged their assistance to the venture and Shackleton was pleased that trustworthy men, such as Wild and Macklin, were returning to his command. When the plans to explore the Arctic fell through, Shackleton merely changed his mission to focus on Antarctica. For Shackleton, in his forty-seventh year, the goals of the expedition were no longer important, just the adventure to nourish his restless spirit.

The *Quest* sailed for the Antarctic on 17 September 1921, but on the voyage southwards Shackleton often complained of tiredness and at one point collapsed. Macklin, the ship's doctor, suspected a heart attack, but Shackleton played down such concerns. On 4 January 1922, the *Quest* approached the mountainous coastline of South Georgia and from the deck Shackleton viewed again the scene of one of his greatest triumphs. In his diary he recorded, 'How familiar the coast seemed as we passed down … It is a strange and curious place. A wonderful evening. In the darkening twilight I saw a lone star hover, gem-like above the bay'.

In the early hours of the next day, Shackleton, in severe pain, called for Macklin. As Macklin gave him medicine Shackleton suffered a massive heart attack and died. He was 47 years old. When Macklin performed a post-mortem, he found evidence that Shackleton had been fighting against heart disease for many years and later remarked that it illustrated Shackleton's 'unyielding determination'. At the request of his wife Emily, Shackleton was buried in the whaler's cemetery at Grytviken, the first whaling station on South Georgia, a fitting home for the restless explorer.

THE MODERN AGE OF EXPLORATION

Shackleton's death marked the end of the Heroic Age of Exploration, where man alone pitted himself against the wilds of nature. The modern age of exploration would be dominated by machines, as technological advances made the white wastes of Antarctica easier to conquer. In 1928, the Australian explorer Hubert Wilkins, became the first person to fly an aeroplane over Antarctica, taking off from Deception Island and flying over the eastern side of the Antarctic Peninsula. This exploit inspired the American pilot Commander Richard Byrd of the US Navy to attempt to fly over the South Pole. Byrd established a base at the Bay of Whales and on 28 November 1929, accompanied by his flying crew, he set off in his Ford Trimotor plane.

> **KEY FACT**
>
> Byrd's Ford Trimotor was specially adapted for Antarctic flight, with high wings and skis instead of wheels for landing on snow or ice.

At the edge of the Ross Ice Shelf Byrd had to clear a 3000-metre mountain range in order to reach the polar plateau – the area of high ground on which the South Pole is located. In the cold thin air the plane struggled to make the steep ascent and the crew were forced to ditch heavy bags of provisions to gain height. The plane cleared the mountain range with only 150 metres to spare and after a four-hour flight reached the South Pole. The hard fought gains that Shackleton had made on foot were now easily bested from the air.

The impact of the Second World War and the lack of new challenges led to a decline in Antarctic exploration. However, in 1957 the **International Geophysical Year**, 12 nations established 50 bases in Antarctica to carry out a range of scientific studies. The British team was led by Dr Vivian Fuchs, whose goal was to achieve Shackleton's unfinished dream of a transantarctic crossing. Instead of the manpower and dogs that Shackleton had planned on using, Fuchs employed aeroplanes to scout the route and travelled across the continent in a convoy of motor-driven tractors. Even with the use of such modern technology Fuchs still found the crossing arduous and, at one point, his tractor fell into a deep crevasse and was rescued only with the aid of metal ramps. Fuchs finally achieved his goal in less than 100 days, after travelling nearly 2,500 kilometres.

> **KEY FACTS**
>
> The International Geophysical Year: a year dedicated to the study of Earth and atmospheric sciences that ran from July 1957 to 31 December 1958.
>
> Antarctic Treaty: an agreement negotiated by the 12 nations present in Antarctica during the International Geophysical Year. It aimed to maintain Antarctica as a place of scientific research and international co-operation.

In 1959 the **Antarctic Treaty** was signed which declared that Antarctica should remain a natural reserve devoted entirely to peaceful research, and military and commercial exploitation of the continent were banned. The treaty also stated that no one nation could lay claim to Antarctic territory, preserving for all the untamed wilds that Shackleton and the other explorers of the Heroic age uncovered.

IN SHACKLETON'S FOOTSTEPS

In recent years, despite the huge advances in equipment and technology, the cycle of Antarctic exploration has turned full circle, as modern-day adventurers have tried to recreate the pioneer spirit established by Shackleton, Scott and Amundsen. In 1992, the British explorer Ranulph Fiennes accompanied by Mike Stroud set out on a march across Antarctica from the Weddell Sea to the Ross Sea coast. Like the explorers of the Heroic Age, Fiennes and Stroud had no aircraft support or technological assistance as they man-hauled their sledges across the Antarctic terrain. It was a tribute to their grit and determination that they completed 2,100 kilometres of their 2,500 kilometre journey before radioing for rescue.

The following year the adventurer Trevor Potts led a team of three men and one woman from Elephant Island to South Georgia in a replica of the *James Caird* named the *Sir Ernest Shackleton*. The wearing journey emphasized for them the scale of Shackleton's feat, achieved as it was without any modern-day assistance or support. More recently, in April 2000, the mountaineers Reinhold Messner, Stephen Venables and Conrad Acker recreated Shackleton, Worsley and Crean's crossing of South Georgia. The experienced mountaineer Messner revealed how difficult he found the crossing even with the aid of modern equipment: 'On the Crean glacier, the situation was so hopeless that we thought we would be unable to get through this ice labyrinth.' After successfully completing the difficult crossing Acker paid tribute to Shackleton's achievement stating, 'The ability to persevere, to keep trying in the face of insurmountable odds, and the tenacity of the human body and spirit are vital lessons that supersede time.'

The desire to relive some of Shackleton's most spectacular achievements is a reflection of his elevated standing in the field of polar exploration. For many years he was overshadowed by the figure of Scott, but more recently Shackleton's abilities as a leader have inspired a new generation of explorers. The swashbuckling Shackleton with his natural affinity for men of all backgrounds and outlooks fits the modern conception of heroism. Unlike Scott and Amundsen, Shackleton did not attain any of his expedition goals but the manner in which he conducted his expeditions, putting the safety and needs of his men before any personal desire for glory, mean that he is an icon not only for explorers but also business leaders who study his leadership skills. In the harsh conditions faced by the *Endurance* expedition, the team-work that Shackleton inspired through his limitless optimism and self-sacrificing behaviour meant the difference between life and death.

The record books may state that Roald Amundsen was the first person to reach the South Pole, but Amundsen himself paid this enduring tribute to Shackleton stating, 'Sir Ernest Shackleton's name will for evermore be engraved with letters of fire in the history of Antarctic exploration. Courage and willpower can make miracles. I know of no better example than what that man has accomplished.'

* * *SUMMARY* * *

- After rescuing the men left on the southern side of South Georgia, Shackleton attempted to rescue the crew stranded on Elephant Island. The thick pack-ice prevented him from reaching the island and he was forced to head to the Falkland Islands for help.

- Shackleton was loaned a tug-boat the *Yelcho*, by the Chilean Navy, which he used to rescue the castaways on Elephant Island. He returned them all to safety without a single crew-member's life being lost.

- When he returned to Britain, Shackleton contributed to the war effort with a propaganda mission to South America and a posting to Murmansk. After the end of the war his restless spirit led him to plan a further Antarctic mission.

- In 1921 Shackleton set off for Antarctica on board his ship, the *Quest*. However, when he reached South Georgia he suffered a massive heart attack and died on 5 January 1922. He was buried in the whaler's cemetery at Grytviken on South Georgia.

- In recent years explorers such as Ranulph Fiennes and Reinhold Messner have attempted to recreate the exploits of Shackleton and other explorers from the Heroic Age of Exploration. Shackleton is now viewed as an inspirational leader, with his techniques and adventures studied not only by explorers, but by business leaders and political figures too.

GLOSSARY

Adze a climbing axe
Bannocks lumps of baked dough
Bo'sun ship's officer in charge of equipment and crew
Bow the front of a ship
Brash a type of ice formed from the wreckage of larger pieces of ice
Calve throw off masses of ice
Chronometer a special type of clock used in navigation
Crevasses deep cracks or holes in the ice
Dead reckoning a method of navigation that works out where a boat's position will be at a certain time, so long as it maintains a specific speed and direction
Depot a place where supplies are kept
Depth perception the ability to distinguish between objects and distances
Dog-driving the use of dogs to pull loaded sledges
Dysentery a form of food poisoning resulting in severe diarrhoea
Elevation height
Fjord narrow inlet of sea
Frostbite an injury to body tissue caused by extreme cold temperature. Severe cases can result in the loss of body parts
Glaciers huge ice rivers that can cut through mountains and form valleys
Hoosh penguin stew
Hummock a pile or ridge of ice
Ice-floe a large sheet of floating ice
Indenture papers a contract committing a person to train as an apprentice
Keel the bottom part of a ship
Leads channels of open water among the pack-ice
Man-hauling when people drag sledges on foot
Manchurian ponies a type of pony found in the mountainous north-eastern area of China
Mate officer on a merchant ship
Meteorologist someone who studies the atmosphere and weather patterns
Pack-ice large floating ice-floes that get pushed together
Parallel a line of latitude
Pemmican strips of dried meat mixed with melted fat
Precipitation rain or snowfall

Pressure ridges places where colliding ice-floes have thrown up huge slabs of sea ice

Primus a portable cooker that burns oil for fuel

Ratings non-commissioned sailors

Sastrugi an adverse surface condition, where the snow has been carved into huge grooves by the wind. Sastrugi ridges are hard as rock and travelling over them on foot or sledge is slow and frustrating

Scurvy a disease caused by the lack of vitamin C which can result in swollen limbs, loss of teeth, depression and, in extreme cases, death

Sextant an optical instrument used for measuring the distance between the sun and the horizon as a means of navigation

Ship's Articles the name given to the contract that sailors agree to uphold when they take service on a ship

Snowblindness a painful swelling of the eyes caused by the glare of sunlight reflected off ice and snow that results in a temporary loss of sight

Spit a point of land stretching into the sea

Sternpost the post at the rear of a ship

Trials test runs

Winter station a base shut down for the winter

FURTHER READING

The Endurance, Caroline Alexander, Bloomsbury, 1999.
The story of the *Endurance* expedition accompanied by Frank Hurley's photographs.

Shipwreck at the Bottom of the World: Shackleton's Amazing Voyage,
    Jennifer Armstrong, Crown Publishing Group, 1998.
A children's book exploring the Endurance expedition.

Shackleton, Kim Heacox, National Geographic Books, 1999.
A retelling of Shackleton's *Endurance* expedition accompanied by lavish photographs.

Antarctica: The Last Continent, Kim Heacox, MapQuest.com, 1999.
An exploration of the wildlife and environment of Antarctica.

Shackleton, Roland Huntford, Abacus, 1996.
A comprehensive account of Shackleton's life and expeditions.

The Last Place on Earth, Roland Huntford, Abacus, 2000.
A comprehensive account of the race to the South Pole between Scott and Amudsen.

Endurance: Shackleton's Incredible Voyage to the Antarctic,
    Alfred Lansing, Carroll and Graf Publishers, 2001.
A vivid retelling of the epic *Endurance* expedition.

Safe Return Doubtful: The Heroic Age of Polar Exploration,
    John Maxtone-Graham, Constable and Company Ltd, 1988.
An exploration of the pioneering days of polar exploration.

Trapped by the Ice: Shackleton's Amazing Antarctic Adventure,
    Michael McCurdy, Walker and Company, 1997.
A children's book recounting Shackleton's *Endurance* expedition accompanied by pictures.

Shackleton's Way: Leadership Lessons from the Great Antarctic Explorer, Margot Morrell and Stephanie Capparell, Nicholas Brealey Publishing Ltd, 2001.
An interpretation of the leadership lessons provided by Shackleton's *Endurance* expedition.

Antarctica, Jeff Rubin, Lonely Planet Publications, 2000.
A comprehensive guide to the Antarctic containing essential information for any visitor.

South, Ernest Shackleton, Constable Robinson, 1998.
Shackleton's personal account of the *Endurance* expedition.

The Heart of the Antarctic, Ernest Shackleton, Signet Book, 2000.
Shackleton's account of the *Nimrod* expedition and his attempt on the South Pole.

Dear Daniel: Greetings from Antarctica, Sara Wheeler, Hodder Children's Books, 2000.
An engaging children's book with travel writer Sara Wheeler describing her experiences of living in Antarctica

Shackleton's Boat Journey, Frank Worsley, W. W. Norton and Company, 1998.
Worsley's own account of the voyage to safety on the ill-fated *Endurance* expedition.

USEFUL ADDRESSES

Scott Polar Research Institute
University of Cambridge
Lensfield Road
Cambridge
CB2 1ER
Telephone: 01223 336540

The oldest international research centre covering the polar regions, the SPRI has the world's most comprehensive polar library and archives, with some of the collection housed in the Shackleton Memorial Library.

The James Caird Society
School Farm
Beneden
Kent
TN17 4EU
Telephone: 01580 240755

An institution committed to preserving the memory of the expeditions of discovery in Antarctica and honouring the leadership skills of Sir Ernest Shackleton.

National Maritime Museum
Greenwich
London
SE10 9NF
Telephone: 020 8312 6632

The National Maritime Museum is dedicated to the history of Britain at sea, and recently hosted a major exhibition charting the history of Antarctic exploration.

INDEX

Adams, Jameson Boyd 38, 41
Amundsen, Roald 1, 44, 84
Antarctic Circle 19
Antarctica
  climate of 3, 4
  seasons of 21, 22, 52
  size of 3
  weather 4–5, 40, 69, 73
Barrier *see* Ross Ice Shelf
Barrier Inlet 33, 35
Bay of Whales 35
Beardmore Glacier 39

Crean, Tom 45, 62, 67, 69, 73–74

Discovery 16, 17, 19, 21
dog-driving 22, 24, 25, 51
dogs 17, 26–27, 28
Dorman, Emily *see* Shackleton, Emily
Dudley Docker 57, 66

Elephant Island 62, 64, 66–72, 77
Endurance 46–9, 50–53, 55, 60

First World War 47, 78
frostbite 40, 41, 64

glaciers 4, 39–40, 41

Hurley, Frank 46, 50
Hut Point 36, 42

James Caird 57, 67, 68, 69, 70, 71, 72

King Edward VII Land 19, 33
King Edward VII Peninsula see *King Edward VII Land*

man-hauling 22, 25, 57, 60–61, 83
Markhams, Sir Clements 15
Marshall, Eric 38, 41
McMurdo Sound 20, 21, 35–36
McNeish, Harry 51, 53, 61, 67
Merchant Navy 13, 14
Mount Erebus 10, 19–21

Nansen, Fridtof 17, 34
Nimrod 34, 35, 36

Ocean Camp 59–60

pack-ice 19, 48, 49, 52, 53, 55, 62, 63
penguins 5–7
ponies 34, 35, 38–9

Quest expedition 80–81

Ross Ice Shelf 11, 19, 33, 82
Ross, James Clark 9–11
Royal Geographical Society 15, 16, 32

sastrugi 25
Scott, Robert
  Discovery expedition 16, 17, 19, 24–30
  tensions between Shackleton and 25, 26, 29, 30, 32, 33
  Terra Nova expedition 44–45
scurvy 27, 28, 29
seals 7, 21
Shackleton, Emily 14, 31, 79, 81
Shackleton, Ernest
  childhood of 12–13
  crossing of South Georgia 73–76
  Discovery expedition 15, 16, 17, 19, 22–23, 24–29, 30

Endurance expedition 1, 47–76
  leadership skills 40, 42, 48, 50, 51, 52, 61, 69, 70, 75, 83–84
  plans for *Nimrod* expedition 32, 34
  plans for *Endurance* expedition 48, 55, 57, 58–60
  Quest expedition 80–81
  rescue from Elephant Island 77–79
  service in Merchant Navy 13–14
  voyage to Elephant Island 62–64
  voyage to South Georgia 67–72
snowblindness 39
South Georgia 67, 68, 70, 71, 72–76, 77, 83
Stancomb Wills 62

Wild, Frank 17, 31
  selection for *Nimrod* expedition 33, 38, 39
  Endurance expedition and 45, 53, 60, 61, 66, 80
Wilson, Edward 16, 24, 25, 26, 27, 32, 33
Worsley, Frank 46
  navigational skills of 48, 63, 64, 67, 68, 69, 70, 71
  Endurance expedition 49, 53, 55, 62, 66, 68–72